T0133608

Pipeline Valve Technology

This book covers the life cycle of pipeline valves, the largest and most essential valves in offshore pipeline engineering. Discussing the design process, testing, production, transportation, installation, and maintenance, the book also covers the risk analysis required to assess the reliability of these valves.

Pipeline valves require particular attention to ensure they are safely designed, installed, and maintained, due to the high stakes. Failure would result in environmental pollution, the destruction of expensive assets, and potential loss of life. Proper installation and upkeep require specialist processes throughout the life cycle of the valve. This book is a key guide to these processes. Beginning by looking at the design of pipeline valves, this book details how conserving weight and space is prioritized, how materials are chosen, how thickness is calculated, and how leakage is minimized. It then discusses production and specific welding techniques to bond dissimilar materials, alongside casting and machining. Building on other discussions in the text with case studies and questions and answers for self-study, this book is the ideal guide to pipeline valves.

This book will be of interest to professionals in the industries of offshore oil and gas, material engineering, coatings, mechanical engineering, and piping. It will also be relevant to students studying coating, welding, mechanical, piping, or petroleum engineering.

Pipeline Valve Technology

A Practical Guide

Karan Sotoodeh

CRC Press is an imprint of the
Taylor & Francis Group, an **informa** business

Seventh edition published 2023
by CRC Press
6000 Broken Sound Parkway NW, Suite 300, Boca Raton, FL 33487-2742

and by CRC Press
4 Park Square, Milton Park, Abingdon, Oxon, OX14 4RN

CRC Press is an imprint of Taylor & Francis Group, LLC

ISBN: 978-1-032-38056-8 (hbk)
ISBN: 978-1-032-38060-5 (pbk)
ISBN: 978-1-003-34331-8 (ebk)
ISBN: 978-1-032-38085-8 (eBook+)

DOI: 10.1201/9781003343318

Typeset in Times
by SPi Technologies India Pvt Ltd (Straive)

Contents

Preface ... xi
Acknowledgment ... xiii
Author Biography ... xv

Chapter 1 Introduction and General Features ... 1

1.1 Introduction ... 1
1.2 General Features ... 2
1.2.1 Valve Types ... 2
1.2.2 Bore Requirement ... 2
1.2.3 Access to the Valve Internals ... 2
1.2.4 Connection to the Pipeline ... 5
1.2.5 Metallic Seats ... 6
1.2.6 Trunnion-Mounted Ball ... 7
1.2.7 Design Standards ... 7
1.2.8 Valve Operator ... 8
1.3 Cross-Section Drawing and Part List ... 9
Questions & Answers ... 12
Bibliography ... 15

Chapter 2 Material Selection ... 17

2.1 Introduction ... 17
2.2 Material Selection ... 18
2.2.1 Body and Bonnet ... 19
2.2.2 Valve Internals (Trim) ... 20
2.2.3 Bolting (Bolts and Nuts) ... 22
2.2.4 Pup Piece and Transition Piece ... 24
2.2.5 Seal ... 25
2.2.6 Stem Bearing ... 26
2.2.7 Spring ... 26
2.3 Coating ... 28
Questions & Answers ... 29
Bibliography ... 30

Chapter 3 Double Isolation and Bleed Concept ... 33

3.1 Introduction ... 33
3.2 Pipeline Valves DIB Concept ... 35
3.3 DIB Effects on the Valve Design ... 38
3.4 DIB2 and Valve Marking ... 43
Questions & Answers ... 44
Bibliography ... 47

Chapter 4 Design for Weight Reduction 49

4.1 Introduction 49
4.2 Design for Weight Reduction 50
4.2.1 Wall Thickness Calculation and Validation 50
4.2.1.1 Wall Thickness Calculation 50
4.2.1.2 Wall Thickness Validation 58
4.2.2 Cylindrical Nuts 65
Questions & Answers 67
Bibliography 69

Chapter 5 Seat Scrapers and Flushing Ports 71

5.1 Introduction 71
5.2 Wax Dissolution in Seats 72
5.2.1 Sealant Injection Fittings 72
5.2.2 Flushing Ports 72
5.2.3 Flushing Fluid 75
5.2.4 Flushing Procedure 75
5.3 Particles Ingress Prevention in Seats 76
Questions & Answers 76
Bibliography 77

Chapter 6 Manufacturing Process 79

6.1 Introduction 79
6.2 Pipeline Valves Manufacturing Processes 79
6.2.1 Kick-off Meeting and Engineering 79
6.2.2 Purchasing or Manufacturing Raw Materials 80
6.2.2.1 Casting 80
6.2.2.2 Forging 82
6.2.3 Welding 83
6.2.4 Machining 84
6.2.5 Inspection and Quality Control 86
6.2.6 Assembly of Valves 87
6.2.6.1 Body and Bonnet Bolts Tightening 87
6.2.6.2 Marking 90
6.2.7 Testing 91
6.2.8 Coating or Painting 92
6.2.9 Preservation and Packing 92
6.2.10 Final Inspection 92
6.2.11 Shipment 93
Questions & Answers 93
Bibliography 96

Chapter 7 Welding Technology....99

7.1 Introduction....99
7.2 Valve End Connections....99
7.2.1 Standard ASME Flange....99
7.2.2 Mechanical Joint (Hubs and Clamps)....99
7.2.3 Compact Flange....100
7.2.4 Wafer Connection....101
7.2.5 Threaded Connection....102
7.2.6 Welded Connections....103
7.2.6.1 Socket Weld....103
7.2.6.2 Butt Weld....104
7.3 A Case Study of Welding Between Pipeline Valves and Two Oil and Gas Export Pipelines....106
7.3.1 Pipeline Material Selection....106
7.3.2 Pup and Transition Pieces....110
7.3.3 Weld Joint NDTs....114
7.3.4 Pipeline Valves Welding....115
7.3.5 Welding Qualification at the Construction Yard....116
Questions & Answers....116
Bibliography....118

Chapter 8 Actuation....121

8.1 Introduction....121
8.2 Actuators' Sources of Power....122
8.2.1 Pneumatic Air....122
8.2.2 Hydraulic....123
8.2.3 Electrical....123
8.3 Actuator Choices for Pipeline Valves....124
8.3.1 Scotch and Yoke Actuator....124
8.3.2 Linear Actuator....125
8.3.3 Electrical Actuator....125
8.4 Actuator Selection for Pipeline Valves....126
8.5 Actuator Selection Case Study....127
Questions & Answers....131
Bibliography....134

Chapter 9 Testing....135

9.1 Introduction....135
9.2 Test Procedure....136
9.3 Test Preparation....136
9.4 Test Setup....137
9.5 Type of Tests....139

9.5.1 Hydrostatic High-Pressure Body or Shell Test......... 139
9.5.1.1 Safety Relief Valve Test......... 141
9.5.2 Hydrostatic High-Pressure Seat or Closure Tests......... 141
9.5.3 Hydrostatic Functional and Torque Measurement Tests......... 143
9.5.4 Hydrostatic Cavity Tests......... 143
9.5.4.1 Both Seats Are SR......... 145
9.5.4.2 Double Isolation and Bleed (DIB1)......... 146
9.5.4.3 Double Isolation and Bleed (DIB2)......... 146
9.5.5 Backseat Test......... 147
9.5.6 Air or Gas Low-Pressure Seat or Closure Tests......... 149
9.5.7 High-Pressure Gas Body or Shell Test......... 149
9.5.8 High-Pressure Gas Seat or Closure Test......... 150
9.5.9 Drift Test......... 150
9.5.10 Electrical Continuity Test......... 151
9.5.11 Fire Test......... 151
9.5.11.1 Fire Test According to API 6FA......... 152
9.5.11.2 Fire Test for Pipeline Valves......... 154
9.5.11.3 Fire Thermal Simulation and Analysis Case Study......... 154
9.6 Actuators Testing......... 157
9.6.1 Hydraulic Actuators Testing......... 157
9.6.2 Electrical Actuators Testing......... 158
Questions & Answers......... 158
Bibliography......... 161

Chapter 10 Preservation and Packing......... 163

10.1 Introduction......... 163
10.2 Preservation Types......... 163
10.2.1 General Preservation......... 163
10.2.2 Internal Preservation......... 166
10.2.3 External Preservation......... 168
10.3 Packing......... 170
Questions & Answers......... 170
Bibliography......... 171

Chapter 11 Handling, Lifting, and Transportation......... 173

11.1 Handling......... 173
11.2 Lifting......... 174
11.3 Transportation......... 176
Questions & Answers......... 180
Bibliography......... 182

Chapter 12 Installation, Operation, and Maintenance....183

12.1 Installation and Operation....183
12.1.1 Storage Before Installation....183
12.1.2 Valve Installation and Operation on the Pipeline....183
12.2 Maintenance....185
12.2.1 Emergency Sealant Injection....185
12.2.1.1 Sealant Injection on Seats....185
12.2.1.2 Sealant Injection on Stem....186
12.2.2 Minor External Leakage Repairs....186
12.2.2.1 Sealant Injection Fitting....188
12.2.2.2 Vent and Drain Blind Flanges....188
12.2.2.3 Modular Valves on the Seat Flushing Ports....189
12.2.3 Major External Leakage Repairs....190
12.2.4 Internal Leakages....192
Questions & Answers....197
Bibliography....199

Chapter 13 Safety and Reliability....201

13.1 Introduction to Shutdown Systems....201
13.2 Safety Integrity Level....202
13.2.1 SIL Case Study for a Pipeline Valve....206
13.3 Failure Mode and Effect Analysis....209
13.3.1 FMEA Case Study for Pipeline Valves....209
13.4 Results and Discussion....212
Questions & Answers....213
Bibliography....214

Index....217

Preface

There are valves everywhere, including in your home's plumbing. In order to drink water in the kitchen, we must open a valve. The same is true to operate offshore pipelines. The most significant valves in offshore pipelines are discussed in this book. If these valves fail, oil and gas will not be able to be transported to customers and the society will be left without enough energy for heating, cooking, and transportation. A reader can be knowledgeable about the most critical valves on the offshore platform, including their design aspects, material selection, failure prevention mechanisms, and environmental protection measures. Pipeline valves are discussed in this book in terms of design, production, testing, coating, preservation, packing, lifting, transportation, and safety considerations. On the offshore platform, a pipeline valve is the largest, heaviest, and most essential valve. A recent project in Norway called for a 38" CL1500 ball valve with a weight of almost 70 kg and a height of over four meters. If pipeline valves fail, many negative consequences may ensue, such as the loss of expensive assets and production, environmental pollution, and the loss of human life. Thus, it is imperative to design these valves accurately with special considerations aimed at minimizing the possibility of leakage into the environment. Furthermore, it is vital to design these valves in such a way as to conserve weight and space on the platforms. Different aspects of design are discussed in this book, including material selection, wall thickness calculations, weight reduction of the body and bonnet, using a double isolation and bleed seat design to reduce seat leakage, using seat scrapers and seat retraction mechanisms, cylindrical nuts, etc. Typically, these valves are welded to the pipeline using special welding techniques to bond two dissimilar materials. In this book, the welding process is explained in detail. The book also covers other aspects of valve manufacturing, such as casting, welding, nondestructive testing, and machining. This book also discusses the importance of valve internal protection during commissioning. Valve testing is an important aspect of ensuring the safety and reliability of the valve. Another topic covered in this book is how to lift and support the valves after installation. Other topics discussed in this book include valve packing, preservation, and transportation. The book will include a few case studies related to valve fire safety design and failure mode and effect analysis (FMEA).

Acknowledgment

Thanks to Flow Control Technology for their assistance in preparing the book as well as the cover image.

Author Biography

Karan Sotoodeh used to work for Baker Hughes as a senior/lead valve and actuator engineer in the subsea oil and gas industry. He got his PhD in safety and reliability in mechanical engineering from the University of Stavanger in 2021. Karan Sotoodeh has almost 16 years of experience in the oil and gas industry mainly with valves, piping, actuators, and material engineering. He has written eight books about piping, valves, and actuators and more than 40 papers in peer-reviewed journals. He has been also selected in international conferences in the USA, Germany, and China to talk about valves, actuators, and piping. Dr. Sotoodeh has worked with many valve suppliers in Europe in countries like the UK, Italy, France, Germany, and Norway. He loves travelling, running, and spending time in nature.

1 Introduction and General Features

1.1 INTRODUCTION

Offshore pipelines are critical elements of the subsea transportation system for delivering hydrocarbon products from offshore platforms, usually to end users and markets. Design, construction, and operation of offshore pipelines are essential tasks. Traditionally installed on offshore platforms, pipeline valves are known as ***riser valves*** in some cases or projects. Figure 1.1 shows three offshore pipeline valves installed on an oil export pipeline. In the picture, these valves are outlined with two equilateral triangles that point toward each other in three cubics. Pipeline valves are vital on an offshore platform because they are the most complex and expensive, the largest and heaviest, and have the longest delivery time. In one instance, three pieces of pipeline ball valves with a 38" diameter and a pressure class of 1500 equivalent to the nominal pressure (250 bar) were used in an offshore oil and gas project, each weighing almost 70 tons and requiring an estimated delivery time of one and half years. This image shows an offshore pipeline in blue that transports crude oil from an offshore platform to users and markets. Offshore pipelines, which contain valves, are critical for the transportation of hydrocarbon products to their destinations. Therefore, offshore pipelines and valves need to be designed, constructed, and operated with a high level of management to avoid malfunctions. This book covers the design, construction, and operation of these valves in detail. The figure shows

FIGURE 1.1 Three pipeline valves located on the oil export pipeline on the offshore platform. (Photo by the author.)

DOI: 10.1201/9781003343318-1

two sections of the pipeline, one of which is underwater or subsea and is typically designed following the Det Norske Veritas (DNV) standard. Another section deals with the platform designed in accordance with the American Society of Mechanical Engineering (ASME) code. The split between the DNV standard and ASME code can affect the design of the third or last pipeline valve. A description of some of the general and key characteristics of offshore pipeline valves is given below.

1.2 GENERAL FEATURES

1.2.1 Valve Types

Pipeline valves used offshore are typically both ball valves and through conduit gate (TCG) valves, and they are used to start and stop fluid flow. Many factors influence the decision between ball and TCG, including cost, delivery time, valve sizes and weights, and client preferences. Pipeline valves are not usually equipped with other valve types such as butterfly valves and wedge-type gate valves due to their high maintenance costs. A second reason why butterfly valves are inappropriate for an export pipeline application is their reduced bore. The following subsection describes the bore requirements of pipeline valves.

1.2.2 Bore Requirement

Reduced bore valves have a smaller internal diameter than the connected (meeting) pipeline, thus they cause greater pressure drops than a valve with the same bore or internal diameter as the pipeline. As a result of the selection and use of reduced bore or port valves for offshore export pipelines, oil production rate and revenue are reduced. Accordingly, pipeline valves should have an internal diameter or bore that is equal to the internal diameter of the pipeline to which they are connected. The second reason to have a specific bore for pipeline valves equal to the internal diameter of the connected pipeline relates to the pipeline inspection gadget (pig) running in the offshore pipeline. A pig is shot into the pipeline (see Figure 1.2) using a pig launcher located on the platform before or upstream of the pipeline valves for various purposes including cleaning and maintenance. As the fluid flows, the pig is pushed through the pipeline. The pig can be used to remove wax deposits from the interior surface of a pipeline as an example of pipeline cleaning. In the course of transportation, crude oil can produce and accumulate wax, which can lead to a reduction in pressure in the pipeline and reduce production. It is recommended that the offshore pipeline valves have the same internal diameter as the connected pipeline, in order to facilitate pig passage within the pipeline. It is common knowledge that pipeline valves must be piggable.

1.2.3 Access to the Valve Internals

Pipeline valves have a ***top-entry*** design, and they are welded to the joining pipe. The top entry represents a design in which the valve internals, such as balls and seats, are accessed from the top of the valve. TCG valves are always top-entry valves, while

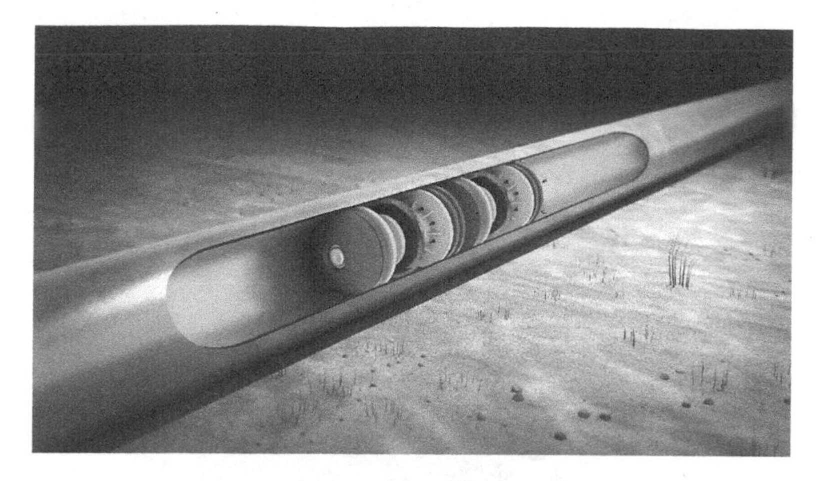

FIGURE 1.2 Pig running in a subsea pipeline. (Courtesy: Shutterstock.)

FIGURE 1.3 A 24″ CL1500 TCG valve. (Photo by the author.)

ball valves can be either top- or side-entry. Figure 1.3 displays a 24″ CL1500 TCG valve, highlighting its body, bonnet, and body and bonnet bolting (bolts and nuts). The most efficient method for accessing the valve interior is to loosen the body and bonnet bolts and nuts. This will enable the operator to remove the valve bonnet from the top of the body. After removing the bonnet from the top of the valve, the internals of the valve can be accessed for repairs or maintenance. Figure 1.4 illustrates the process of removing a stem and disk from a TCG after removing the body and bonnet

FIGURE 1.4 Removing the bonnet, stem, and disk from the valve internal. (Photo by the author.)

bolts and lifting the bonnet. In Figure 1.5, the ball valve is shown during the factory acceptance test (FAT) with its bolted body and bonnet (cover).

In the case of a ***side-entry*** valve, also called a ***split body***, two or three body pieces are connected by bolts. In addition, nuts are assembled and disassembled by the valve manufacturer for maintenance or any other reason from the side. As illustrated in Figure 1.6, a side-entry ball valve includes two body pieces that are bolted together as two flanges. As shown in the figure, there is no access to the valve from the top. In this instance, the valve internals are accessed by unscrewing the bolts that secure the body pieces to each other and separating the second body piece from the first.

Inline maintenance is the primary reason and advantage of the top-entry design for pipeline valves. The valves may be maintained more easily without having to disassemble them from the pipeline. Another advantage of the top-entry design is its higher resistance to pipeline loads due to its one-piece construction. Moreover, a two- or three-piece side-entry valve body is more susceptible to leaks than a top-entry valve. A top-entry valve is welded from both sides to the pipeline, in contrast to a side-entry valve with flange connections. The possibility of leakage from the welded joint is less than that from a flange connection. Consequently, top-entry designs are less likely to leak than side-entry designs due to their welding joint with the pipeline rather than the flange connection. Last but not least, the top-entry design offers greater flexibility in stem enlargement due to high actuator forces. The disadvantages of the top-entry design include higher costs, longer delivery times, and heavier weight when compared to the side-entry design.

FIGURE 1.5 The top-entry ball valve during FAT. (Photo by the author.)

FIGURE 1.6 Ball valve with a side entry and two pieces bolted together. (Photo by the author.)

1.2.4 Connection to the Pipeline

Pipeline valves are welded to the connected pipeline rather than being connected by flanges. An advantage of a flange-end connection for a valve is that it can be disassembled from the connected pipe through its flange connections for routine maintenance, inspection, or cleaning. In contrast, top-entry pipeline valves can be maintained inline without having to be removed from the pipeline. In addition, welded joints are less likely to leak than flange connections. The body of the valve is not directly welded to the pipeline because the heat of welding can damage soft or non-metallic seals inside the valve. Valve seals prevent fluid leakage from valve parts

FIGURE 1.7 A 38" valve body, a transition piece, and a pup piece welded together. (Photo by the author.)

either internally or externally. The damage to valve seals has adverse consequences, including inefficient valve and system operation, health, safety, and environmental issues (HSE). ***Pup pieces*** are two pieces of pipe attached to the pipeline valves on one side and to the pipeline itself on the other side, in order to prevent damage to the seals of the valves due to welding to the pipe in the yard. It may be necessary to weld a ***transition piece*** between the pup piece and the valve's body. This concept is discussed in more detail in Chapter 7. Figure 1.7 illustrates a welding shop where workers apply welding to the body and pup pieces of the valve. This photograph shows the body of a 38" CL1500 (pressure nominal of 250 bar), welding between the valve's body and the transition piece, and welding between the transition and the pup piece. The same transition and pup piece arrangement exists on the other side of the valve.

There is no direct connection between side-entry valves and piping; a flange is welded to the piping from one side and bolted to the valve body from the other side. One drawback of the side-entry ball valve is the need to add double flanges on both sides in order to connect the valve ends to the piping. In contrast, top-entry ball valves do not require these two flanges, which add weight, space, and cost to piping procurement and construction.

1.2.5 Metallic Seats

It is possible for valve seats to be soft (non-metallic), such as thermoplastic materials (e.g., polyether ether ketone), or metallic. Thermoplastics are made from polymers, which become soft when heated and stiff when cooled. Soft seats provide the advantages of a tight seal, ***bubble-tight***, or ***zero leakage***. In addition, soft-seated valves are less expensive than valves with metallic seats. However, soft-seated valves are not suitable for particle-containing services, high-temperature applications, or situations in which the valve is frequently operated or employed under high differential pressure between upstream (before) and downstream (after). The pipeline, including the valves, is subject to pig running, which can damage the soft seat. Moreover, pipeline valves may be operated against high differential pressure, so metal seats are recommended for them according to the author's experience in industrial projects.

1.2.6 Trunnion-Mounted Ball

Ball valves can be ***floating*** or ***trunnion mounted***. Between these two designs, the most significant difference is the construction of the ball and the way it is assembled in the valve. A trunnion design supports the ball below in the form of a plate or flange called a "trunnion." Therefore, the ball is secured between the stem from the top and the trunnion from the bottom of the trunnion design. The trunnion design indicates that the ball is not floating but fixed and centered, making it the ideal choice for large and high-pressure pipeline valves. As opposed to the trunnion-mounted ball, a floating ball is only connected to the stem from the top and has greater ability to move within the body of the valve as a result of the inline fluid pressure. Although trunnion-mounted designs are more expensive than floating designs, they are safer, and the amount of force required to operate a trunnion-mounted ball valve is generally lower than that of a floating ball valve of the same size and pressure class. The floating ball design, as opposed to the trunnion design, is more typical for small-sized and low-pressure class ball valves (e.g., 2" CL150). Figure 1.8 shows a floating ball valve on the left and a trunnion ball valve on the right.

1.2.7 Design Standards

While engineering contractor companies typically specify API 6D, Specification for pipeline and piping valves for offshore pipeline valves in engineering documents,

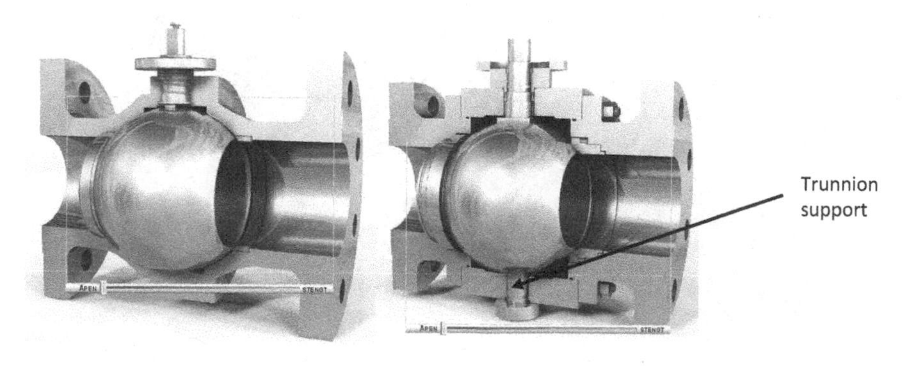

FIGURE 1.8 Comparison of a floating ball valve (left) versus a trunnion ball valve (right). (Photo by the author.)

including piping and valve material specifications and datasheets for the valves, manufacturers' standards, and norms apply to the design and manufacturing of these valves. It is clear that, for example, the internal diameter and end-to-end dimensions of the pipeline valves do not comply with API 6D requirements, but rather follow the client's and manufacturer's specifications.

1.2.8 Valve Operator

Valves are operated manually by levers or handwheels and gearboxes, or automatically by actuators. The manual operation does not apply to pipeline valves because it takes a very long time for an operator to operate (open and close or vice versa) a pipeline valve owing to its large size and high-pressure class. An operator may need approximately half an hour to manually open or close a pipeline valve. Some pipeline valves have an emergency shutdown (ESD) function to be closed in a very short time (e.g., less than one second for a one-inch size valve), which cannot be achieved by manual operation certainly. Hydraulic and electrical actuators are popular choices for pipeline valves. Electrical actuators are selected to provide only ease of operation for pipeline valves, whereas hydraulic actuators are used to facilitate the ESD function of the valve. Detailed information about pipeline valve actuation can be found in Chapter 8. Figure 1.9 illustrates two pipeline ball valves with the electrical actuators installed on the top of the valves on the right and a ball valve on the left with a hydraulic actuator in red on its top.

FIGURE 1.9 Three pipeline valves, one with a hydraulic actuator on the left and two with electrical actuators on the right. (Courtesy: Flow Control Technology.)

1.3 CROSS-SECTION DRAWING AND PART LIST

A cross-sectional drawing demonstrates what an equipment or component (such as a valve) looks like, including all its components and selected materials. An offshore pipeline valve is shown in cross-section in Figure 1.10. In this section, some

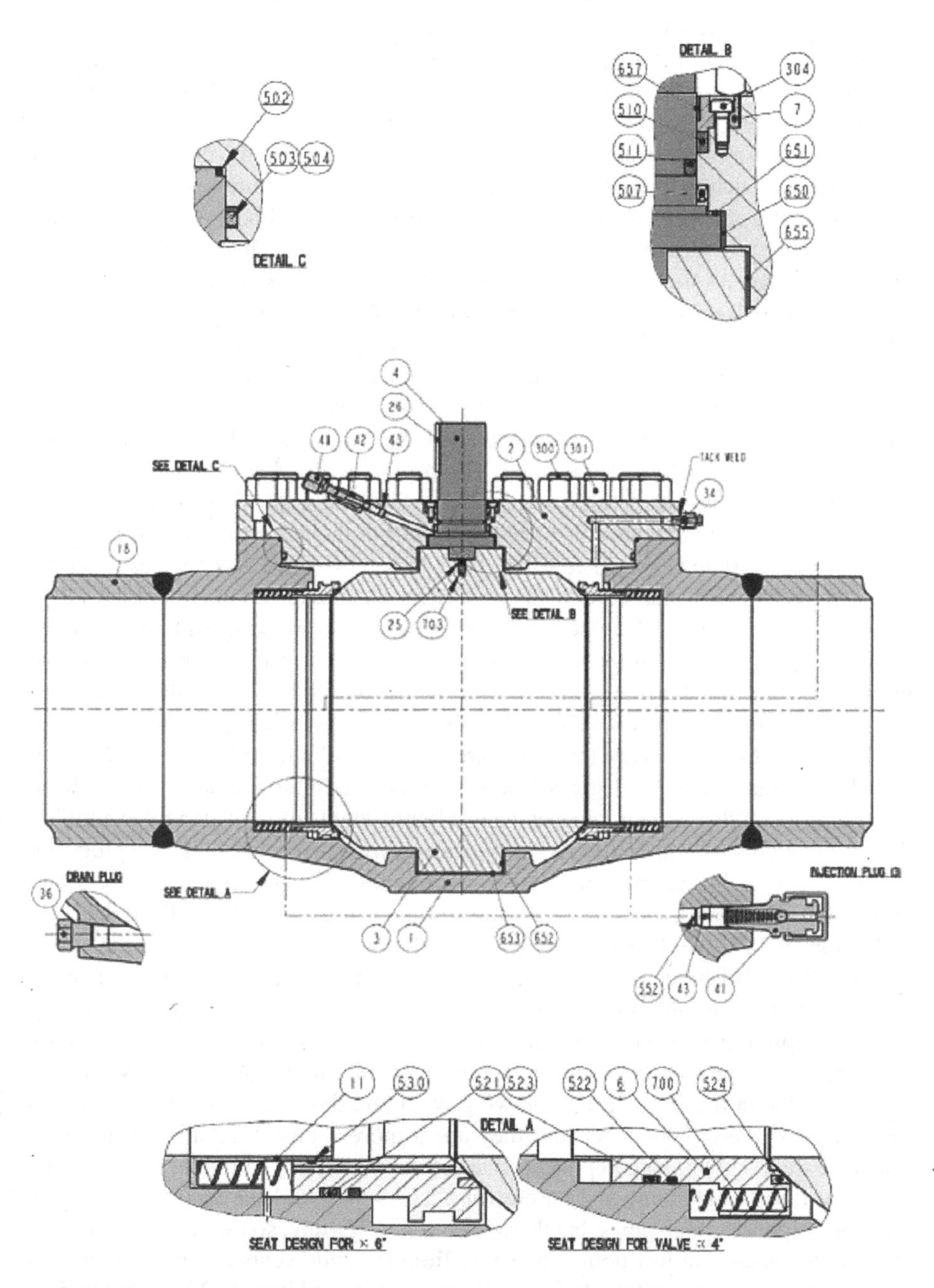

FIGURE 1.10 An illustration of a pipeline valve cross-section.

information will be provided regarding the various parts of this valve. In the picture above, items #1 and #2 are the ***body*** and ***bonnet*** of the valve, respectively. The body of the valve is the outer casing that houses the internal components. The body is a pressure-contained component, which if not functioning properly will result in the release of fluid contained within into the environment. Generally, the bonnet or cover is screwed or welded into the body from the top and serves as the valve's body cover. The third component is the ***ball***, also known as the closure member, which controls the flow rate. In item #4, there is a ***stem*** that connects to the valve operator (e.g., a gearbox) from the top and the valve closure member from the bottom, providing the required force for opening and closing the valve. In the valve body, item #6 is the ***seat*** or ***seat ring***, which is in contact with the ball to prevent internal leakage. The item #7 is the ***cover***, and as its name suggests, it covers the entire body. As shown in item #700, a ***spring*** provides the necessary force to push the seat against the ball to prevent internal leakage. As the spring moves, the spring is guided by item #11, also known as the ***spring guide***. In item #18, there is a ***pup piece*** welded to the valve's body from one side, and Section 1.2.4 describes the pup piece in more detail. ***Antistatic ball*** number 25 and antistatic spring number 703 are located between the ball and stem of the valve. They provide electrical continuity between these parts and to facilitate discharging of the static electricity produced inside the valve due to friction between the valve's components while in operation. The production and accumulation of static electricity inside the valve can cause a spark, fire, and explosion. As a result, the ***antistatic spring*** and ball are considered to be part of the valve fire-safe design. Item #26 is a ***shaft*** or ***stem key*** coupled with a stem. When the valve is connected to piping and is in operation, a gearbox or actuator may be removed from the top of the valve. It is impossible for the valve operator or other personnel working with the valves to know the exact position of the valve in this situation. As a consequence, one or two grooves are machined inside the valve stem, where one or two bars called stem keys are installed. In quarter-turn valves, such as ball valves, stem keys are commonly used. Quarter-turn valves are those that are opened and closed by turning the stem and closure member by 90 degrees. The key(s) are parallel to the hole of the ball, so the operator can distinguish the opening or closing status of the valve by examining the key(s). Items 34 and 36 are ***vent and drain plugs***, respectively. These plugs are installed in the top and bottom of the valve cavities and are removed to vent and drain the trapped fluid. Valve cavities refer to the area between the ball and the body and bonnet of a valve. During the opening and closing of the ball valve, the cavity can be filled with fluid. Fluids inside the cavity can be pressurized and expanded and damage the ball, the body, the bonnet, and even the stem. Therefore, the use of vent and drain plugs or flanges on the valve cavities is essential to utilizing a cavity pressure release or relief mechanism. While vent and drain plugs are shown for the pipeline valve in the cross-section drawing, vent and drain flanges are more typical for pipeline valves. This is because they are a more robust alternative than plugs. Plugs are threaded piping components that are susceptible to galling or metal sliding and wear during tightening and unscrewing. The vent and drain flange connections are more expensive than plugs, but there is no risk of galling, so they are more secure. Drains are installed at low points to release liquids, while vents are installed at high points to release gases. Item #41 is an ***injection plug***, and item #42 is its ***extension***.

Installed on the stem and seat areas, the injection plug is known as a ***sealant injection fitting***. The stem of the valve is sealed with soft (non-metallic) seal materials, such as O-rings and lip seals, to prevent leakage and emissions to the environment. Since the stem seals can be damaged during valve operation, the sealant is injected through the sealant injection fitting for emergency repair of the seals. There is a small ***check valve*** #43 on the ***stem sealant injection fitting*** to prevent backflow of sealant in the injection system. The 300 and 301 items are the body and bonnet-connected stud bolts and nuts, respectively, used to secure the valve's body and bonnet together.

Item #304 is a ***screw*** used to connect the cover bonnet to the body. A couple of seals are located between the body and bonnet; item #503 is an ***O-ring***, and item #502 is an extra ***graphite*** or ***fire-safe ring***. In the event of fire, the O-ring is melted, so the graphite ring preserves the seal between the valve body and bonnet. Items #507 and #511 are stem seals or ***lip seals***, which consist of a metallic spring that exerts force on a non-metallic material (e.g., Teflon) to provide sealing. Lip seals are melted during a fire, and they are not able to seal, so a fire-safe ring in graphite, item #510, is used for stem sealing during a fire. Items #521 and #523 are two lip seals, also called seats to body seals, located between the valve's seats and its housing to prevent leakage from these areas. During a fire, the lip seals between seats and the valve's body are destroyed, so a couple of fire-safe graphite rings (Items #522) are designed as backups. Item #524 is a ***seat insert*** made from a soft material like ***polyether ether ketone (PEEK)***. However, offshore pipeline valves are metal seated without a seat insert. Item #530 is known as a ***seat scraper***. A seat scraper is a protective component used to remove particles, dirt, or debris that may damage or clog the seat. There are two ***bearings*** with the part numbers 650 and 657, one lower and one upper. The bearings are installed around the stem in order to prevent the stem from moving sideways due to the applied loads from the valve operator. Items #652 and #655 are lower and upper ball bearings, which are designed to prevent side-to-side movement of the ball. Items #651 and #653 are thrust bearings or washers installed to prevent the stem or ball from drifting or blowing out in the axial direction. Table 1.1 lists the valve's parts and quantities.

TABLE 1.1
Part List of the Pipeline Valve in Figure 1.10

Item Number	Designation	Quantity
1	Body	1
2	Bonnet	1
3	Ball	1
4	Stem	1
6	Seat ring or seat	2
7	Cover	1
18	Pup piece	2
25	Antistatic ball	1
26	Stem key or shaft key	1
34	Vent plug	1
36	Drain plug	1

(Continued)

TABLE 1.1 (CONTINUED)
Part List of the Pipeline Valve in Figure 1.10

Item Number	Designation	Quantity
41	Injection plug	3
42	Injection plug extension	3
43	Mini check valve	3
300	Body stud bolt	*
301	Body nut	*
502	Fire safe seal (graphite)	2
503	O-ring gasket	2
507	Stem seal (lip seal)	2
510	Fire safe seal (graphite)	2
511	Stem seal (lip seal)	2
521	Seat to body seal (lip seal)	1
522	Fire safe seal (graphite)	2
523	Seat to body seal (lip seal)	1
524	Seat insert	2
530	Seat scraper	2
650	Lower stem bearing	1
651	Thrust washer	1
652	Lower ball bearing	1
653	Thrust washer	1
655	Upper ball bearing	1
657	Upper ball bearing	1
700	Spring	*
703	Antistatic spring	1

* It depends on the valve size and pressure class.

QUESTIONS & ANSWERS

1. Identify the item that does not describe the characteristics of an offshore pipeline valve.

   A. Heavyweight
   B. High-pressure class
   C. Designed as per ASME B31.3
   D. Piggability

   Answer) Option C does not describe the characteristics of an offshore pipeline valve. Pipeline valves for offshore pipelines typically comply with API 6D, Specification for piping and pipeline valves, as well as manufacturer standards and client requirements. ASME B31.3 is a code of requirements for the materials, design, fabrication, assembly, erection, examination, inspection, and testing of the process piping.

2. The internal diameter of an offshore pipeline is 849 millimeters. Figure 1.11 illustrates the positioning of a pig in a pipeline with a minimum diameter or height, H, equal to 97% of the diameter of the pipeline. What could be the internal diameter of the valve attached to the pipeline?

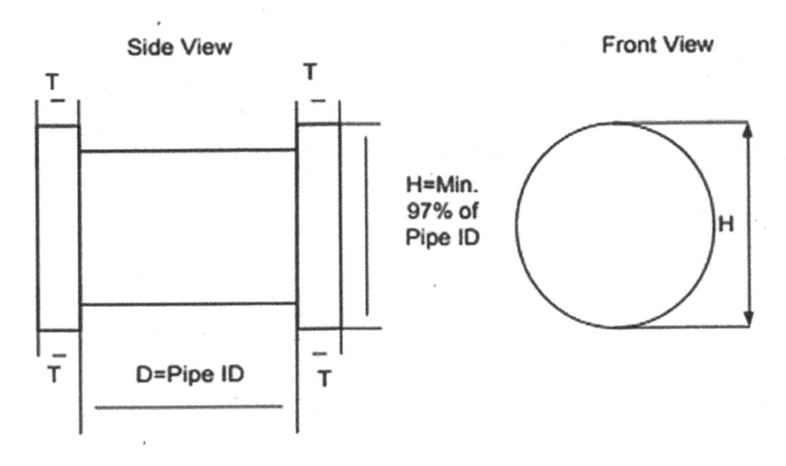

FIGURE 1.11 A sketch of the offshore pipeline pig.

A. 849 mm
B. 829 mm
C. 850 mm
D. 840 mm

Answer) A pig whose diameter exceeds the inner diameter (ID) of a pipeline will not be able to pass through it. Pig heights and diameters may be undersized (e.g., 97% of the pipeline's ID in this case) to prevent excessive wear between the pig and the pipeline's inner surface as well as damage to the pig.

In general, pipeline valves have the same internal diameter as the pipeline they connect to, in this case 849 mm. Nevertheless, the inner diameter of pipeline valves may be affected by the diameter or height of the pig. The valve designer can, for example, reduce the internal diameter of the pipeline valve to a few millimeters over the minimum height of the pig. In this case, the minimum height of the pig is calculated as 97% * 849 (Pipeline ID), which is 823 mm. Providing a clearance of 6 mm, 829 mm is the closest choice. Accordingly, option B is the correct answer.

3. What is the primary advantage of the top-entry design over the side-entry design?

A. Lower cost
B. Inline maintenance possibility
C. Lower weight
D. Shorter delivery time

Answer) Option B is the correct answer.

4. Figure 1.12 shows two ball valves, one with a top entry on the left and one with a split or side entry on the right. Which statement is true about their designs?

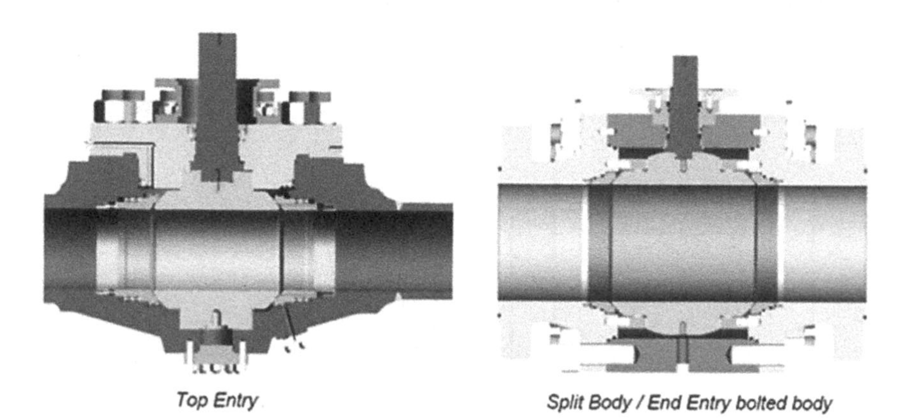

FIGURE 1.12 Designing a top entry vs. a side entry.

A. The side-entry ball valve is leakage-free.
B. The top-entry ball valve is connected to the pipeline by flanges.
C. The side-entry valve can be maintained inline.
D. In comparison to the side-entry design, the top-entry design has more mechanical strength.

Answer) Option A is not the correct answer, as it is possible for leakage to occur from the end flange connections and body pieces, meaning that its design is not leakage-proof. Option B is not appropriate either since the top-entry design valve is welded to the pipe. Option C is incorrect since inline maintenance only applies to top-entry designs, not side-entry designs. Therefore, Option D should be selected.

5. Which sentence is correct about pipeline valves operation?

A. Pipeline valves are operated manually.
B. Electrical actuators are suitable choices for the valves with ESD function.
C. Pneumatic actuators are generally not selected for pipeline valves.
D. Pipeline valves always require a speedy operation of less than one second per inch size.

Answer) Option C is the right answer. Pipeline valves have actuators and aren't manual, so option A isn't right. Electrical actuators aren't suitable for ESD functions because they have a much slower speed of operation than hydraulic actuators. Option D is wrong because pipeline valves need to move quickly.

BIBLIOGRAPHY

1. American Petroleum Institute (API) 6D. (2014). *Specification for pipeline and piping valves* (24th edition). Washington, DC: API.
2. Amyotte P. & Faisal Khan (2020). *Methods in chemical process safety*. Elsevier Science. ISBN: 012821824X, 9780128218242.
3. Sotoodeh K. (2018). Why are butterfly valves a good alternative to ball valves for utility services in the offshore industry? *American Journal of Industrial Engineering*, 5(1), 36–40. https://doi.org/10.12691/ajie-5-1-6
4. Sotoodeh K. (2021). Dissimilar welding between piping and valves in the offshore oil and gas industry. *Welding International*. Taylor & Francis. https://doi.org/10.1080/09507116.2021.1919495

2 Material Selection

2.1 INTRODUCTION

The selection of material is a critical step in the design of products such as piping and valves. The selected materials should not fail due to corrosion, loads, erosion, etc. during the product's design life (e.g., 20 or 25 years). Material selection becomes more complex when several technical parameters are taken into consideration, such as corrosion, erosion, operating temperatures, mechanical strength, and weight. Furthermore, the non-technical parameter of cost plays an imperative role in the material selection process. As shown in Diagram 2.1, there are several vital parameters and aspects to consider when selecting materials for piping and valves in the offshore oil and gas sector.

In this section, we present some considerations for selecting materials for industrial valves used in offshore applications, such as:

- Type of fluid passing through the valve, and its corrosiveness by corrosive compounds like chloride, carbon dioxide (CO_2, and H_2S)
- Fluid service pressure and temperature
- Possibility of galvanic corrosion because of contact between two dissimilar materials in an electrolyte
- Crevice corrosion in the grooves (sealing areas) of non-corrosion-resistant materials such as carbon and low-alloy steels
- Overlay welding of corrosion-resistant alloys (CRAs) on carbon and low-alloy steels
- Material availability and cost
- Non-metallic sealing materials and their compatibility with operating temperatures and chemical fluids
- The design life of the valves
- Coating selection and requirements for external corrosion protection of the materials
- Compatibility of materials with injected fluids

Galvanic corrosion, also called dissimilar metal corrosion or electrolytic corrosion, occurs when a metal or alloy is electrically coupled with another metal or non-metal in the same electrolyte. As dissimilar metals are electrically coupled and form galvanic coupling, the corrosion of the metal with less corrosion resistance or higher activity increases, becoming an anode, whereas the corrosion of the metal with more corrosion resistance decreases, becoming a cathode. Corrosion between these dissimilar materials is driven by the difference in electrical potential between them.

Crevice corrosion is a type of corrosion that takes place in confined spaces where fluid flow is limited. Crevice corrosion sometimes occurs before pitting corrosion starts, and it often has a very similar mechanism to that of pitting corrosion. Crevices

DOI: 10.1201/9781003343318-2

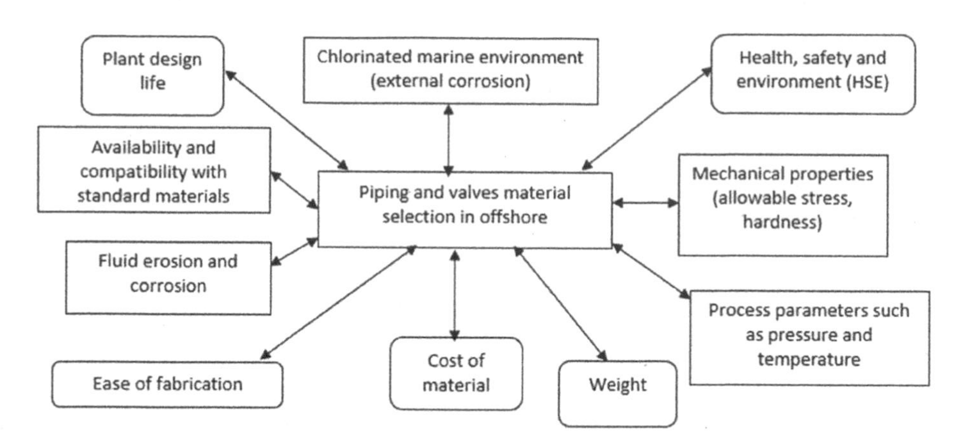

DIAGRAM 2.1 Key factors in material selection for piping and valves in the offshore oil and gas industry. (Chart by the author.)

can develop under gaskets, seals, and washers. Crevice corrosion can occur in carbon steel, stainless steel (SS), nickel alloys, aluminum alloys, titanium alloys, and copper alloys; specifically, it is severe in carbon and low-alloy steels.

2.2 MATERIAL SELECTION

This section provides further details regarding material selection philosophy for offshore pipeline valve components. They include the body and bonnet, internals such as the ball and seat, stem, pup, and transition pieces, as well as seals. A few of the valve's components are shown in Figure 2.1.

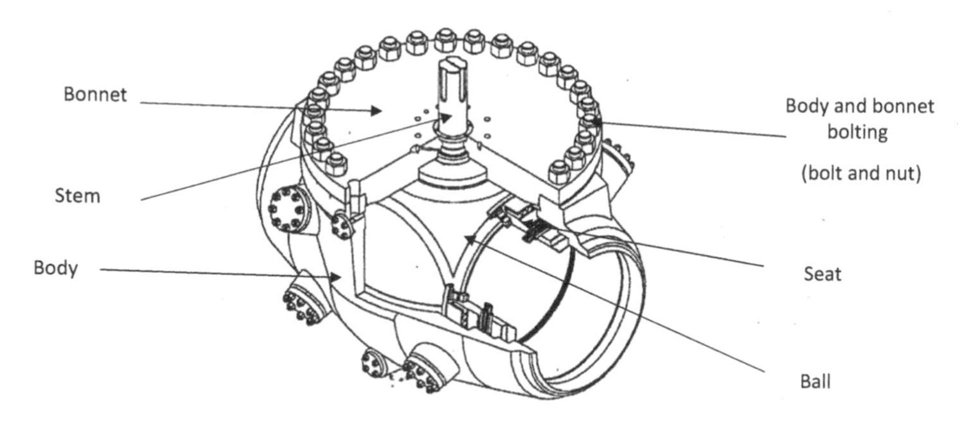

FIGURE 2.1 A top-entry offshore pipeline valve including its parts. (Photo by the author.)

2.2.1 Body and Bonnet

A valve's material selection usually begins with the body or body and bonnet. The offshore pipeline valves which handle oil and gas fluids are installed on oil and gas export pipelines. Prior to this, upstream facilities treated pipelines transporting oil and gas fluids to remove corrosive compounds such as carbon dioxide and hydrogen sulfide. Thus, CRAs are not required for the body and bonnet of export pipeline valves. CRAs are SSs, nickel alloys, titanium, and other metal alloys that contain chromium, nickel, molybdenum, etc. Note that carbon steel and low-temperature carbon steel (LTCS) are not considered CRAs, but they are frequently used for the body and bonnet of pipeline valves.

Carbon steel, also called steel, is a non-CRA consisting of carbon and iron compounds, and as the name suggests, has a larger concentration of carbon than other elements. The maximum carbon content of carbon steel is two percent. Carbon steel is divided into four categories: low, mild, medium, and high-carbon steel, depending on the carbon concentration. Carbon steel can be low carbon from 0% to 0.15%, mild carbon from 0.15% to 0.3%, medium carbon from 0.3% to 0.5%, and high carbon from 0.5% to 2%. A carbon–iron alloy with more than 2% carbon content is called cast iron. Carbon greatly influences the mechanical properties of carbon steel, including its mechanical strength, hardness, and ductility. Increasing the carbon content increases the mechanical strength and hardness but decreases the ductility and weldability of carbon steel. Typically, high and medium carbon steels are not commonly used for piping and valves in the oil and gas industry. The reason for this is their higher brittleness and poorer weldability in comparison to low carbon content carbon steel.

Carbon steel can be used in design temperature ranges from –29°C to approximately 400°C as per the American Society of Mechanical Engineers (ASME) B31.3, process piping code. The operating temperature of the pipeline valves does not exceed 150°C, but the minimum design temperature of these valves could be less than –29°C. The LTCS is chosen for low-temperature pipeline valve bodies and bonnets if the design temperature is between –29°C and –46°C. The minimum design temperature of the pipeline valves does not reduce to a value less than –46°C. Although carbon steel and LTCS are very similar in chemical composition, LTCS has minimal to medium carbon and a higher amount of manganese than carbon steel, improving its toughness resistance in low-temperature applications. Material is tough if it can withstand high stress and deformation before fracturing. Among the failure mechanisms of low temperatures are fracturing and embrittlement. In addition to minor chemical composition differences, a ***Charpy impact test*** is performed on LTCS at a low temperature of –46°C to ensure that the material has enough impact toughness or strength and can function properly at this temperature. A Charpy impact test or Charpy V-notch test is performed on different materials to measure their impact toughness or strength through the required impact energy absorbed by the materials during fracturing. Impact strength or toughness represents the ability of the material to resist fracture by a blow. The higher impact energy absorption capacity of material before fracture means higher toughness.

The American Society for Testing Materials (ASTM) is an international standards organization founded in 1902 that classifies various steels by their composition and physical properties. ASTM standard covers different materials such as iron and steel, non-ferrous metallic, ceramic and concrete, and miscellaneous materials. Iron and steel materials like carbon, low-alloy steel, and SS typically start with the letter "A." The valves' carbon steel body and bonnet can be forged or cast. The forged carbon steel body and bonnet are designated by ASTM A105, while the cast carbon steel body and bonnet are shown by ASTM A216 WCB. If the valve has an LTCS body and bonnet, ASTM A350 LF2 is the forged LTCS designation, and ASTM A352 LCB or LCC is the cast material designation. Figure 2.2 illustrates the cast body of a pipeline valve in LTCS and ASTM A352LCC. The advantage of LCC over LCB is that it contains less carbon and more manganese, which makes it more weldable. The pipeline valve body is welded to the transition or pup pieces on both sides. Figure 2.3 illustrates the bonnet of the same pipeline valve in forged LTCS, ASTM A350 LF2.

There are groove-shaped areas on the pipeline valve's body and bonnet where seals or gasket are inserted. Seals and gaskets between two joints prevent internal fluid leakage (e.g., oil and gas). Internal fluid can be trapped inside the areas beneath the seals and gaskets and cause crevice corrosion. The crevice corrosion is more severe for non-CRAs like carbon and low-alloy steels and LTCS than CRAs. Therefore, 3-mm Inconel 625 is weld-overlaid on the seat pockets and seal areas to prevent crevice corrosion. Table 2.1 summarizes offshore pipeline valves' body and bonnet materials.

2.2.2 Valve Internals (Trim)

Valve internals or ***trim*** include components such as the closure member (e.g., disk or ball), seats, and stem in direct contact with the internal fluid. According to API standards, trim materials must be at least as corrosion resistant as the body materials. However, carbon steel trim and internals for a carbon steel body and bonnet valve are an improper choice of material that results in corrosion, erosion, and failure of the valve's ball, seats, and stem. Consequently, 13Chromium-4 Nickel martensitic SS is a popular material for all trim parts of pipeline valves. Martensitic SS, also known as

FIGURE 2.2 The cast body of a pipeline valve in LTCS according to ASTM A352LCC. (Photo by the author.)

FIGURE 2.3 A pipeline valve bonnet in forged LTCS, ASTM A350LF2. (Photo by the author.)

TABLE 2.1
Details of the Body and Bonnet Materials

Body Base Material	Body Material Grade Detail	Bonnet Material Grade Detail
Carbon steel	ASTM A216 WCC + Inconel 625 partially overlaid on the grooves	ASTM A105 + Inconel 625 partially overlaid on the grooves
LTCS	ASTM A352LCC + Inconel 625 partially overlaid on the grooves	ASTM A352LCC + Inconel 625 partially overlaid on the grooves

series 400, has a chromium content ranging from 12% to 17% and a carbon content ranging from 0.10% to 1.2%. Martensitic SS is known for its hardness and mechanical strength. Harder materials are more resistant to erosion. Since the ball and seat of the valve are constantly in contact with the fluid, they must be highly corrosion and erosion resistant. Moreover, during the opening and closing of the valves, metal-to-metal friction occurs between the balls and seats. Wear and galling between the contact surfaces of the ball and the seat are a result of surface adhesion of the sliding surfaces. Hardfacing alloys such as tungsten carbide are applied to the ball and seat contact surfaces to prevent galling and wear. Hardfacing with tungsten carbide provides the highest degree of abrasion resistance.

Because the actuators of industrial valves place stress on the stems, the stems must be highly mechanically strong. Hydraulically actuated valves should have a more robust stem due to the higher loads applied by the actuators. Because of its high mechanical strength, 13Chromium-4 Nickel (13Cr-4Ni) martensitic SS is a suitable material for offshore pipeline valve stems. In the oil and gas industry, 13Cr-4Ni SS

FIGURE 2.4 The ball and stem of a pipeline valve made from 13Cr-4Ni forged to ASTM A182F6NM. (Photo by the author.)

(UNS S41500) has been used to replace 13Cr (UNS S41000) due to its superior mechanical properties, toughness, low temperature resistance, and corrosion resistance. For the internals of pipeline valves, this alloy can either be forged or cast. A forging of 13Cr-4Ni is known as ASTM A182 F6NM, and the cast is designated as CA6NM according to ASTM A487, ASTM A352, and ASTM A743. Forged stems are more common in industrial valves. Figure 2.4 illustrates the ball and stem of an offshore pipeline valve made of forged 13Cr-4Ni, ASTM A182F6NM. It is noteworthy that the ball has been hardfaced with tungsten carbide in order to prevent erosion due to contact with metallic seats during valve operation.

The martensitic SS 17-4 PH offers higher mechanical strength than the 13Cr-4Ni martensitic SS. In onshore plants, such as refineries and petrochemical plants, 17-4PH is a popular choice of stem material for carbon steel body valves. 17-4PH is, however, easily corroded by the corrosive offshore environment, so it is not an acceptable choice for offshore carbon steel body valves. 17-4 PH, also known as UNS S17400 or grade 630, is a martensitic SS with exceptional mechanical strength and hardness due to precipitation hardening. In precipitation hardening, also known as age hardening or particle hardening, extremely tiny particles, called the second phase, such as aluminum or copper, are uniformly dispersed in the original material to increase its strength and hardness. Second-phase copper is used in 17-4 PH to achieve high mechanical strength. This material is known as 17-4 PH because it contains 17% chromium and 4% nickel. There is also 4% copper in this material.

2.2.3 Bolting (Bolts and Nuts)

Bolting refers to the use of bolts and nuts to connect valve parts. In offshore pipeline valves, the body and bonnet are held together with bolts and nuts. Figure 2.5 illustrates a 38" ball valve with a pressure class of 1500 equal to 250 bar, ready for transportation and installation on an oil export pipeline. Bolts and nuts connecting the body to the bonnet are highlighted in the photograph. When these bolts and nuts

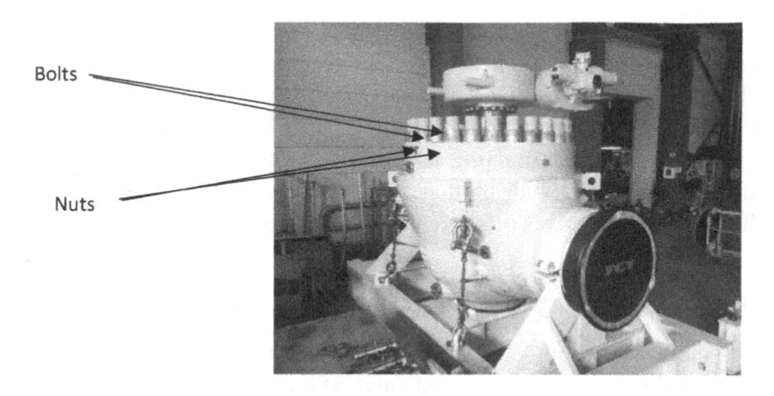

FIGURE 2.5 A 38" CL1500 ball valve is shown with the bolts and nuts that connect the body and bonnet of the valve.

fail to hold the body and bonnet of the valve together, oil leaks from the valve. As a result, negative consequences such as pollution and product loss may occur, as well as the possibility of fire and exposure.

Low-alloy steel bolts and carbon steel nuts are widely used for carbon steel body valves in different sectors of the oil and gas industry, such as offshore, refineries, and petrochemical plants. Low-alloy steel bolts are manufactured from alloys containing chromium and molybdenum that are compatible with AISI 4140 and are quenched and tempered to provide high mechanical strength. In quenching, a material is heated to a certain temperature (e.g., 900°C) and then cooled in water or oil to complete the hardening process. Following this, tempering is accomplished by heating the metal to a high temperature below its melting point. This is done by cooling it, usually in the air, to improve material characteristics such as toughness and reduce brittleness. AISI is a material standard used to specify specific standard carbon and low-alloy steel grades based on four-digit codes. Low-alloy steel bolts are not commonly coded according to AISI. Instead, ASTM codes are applied. A193 B7 bolts are a widely used grade of low-alloy steel bolts. In conjunction with ASTM A193 B7 bolts, ASTM A194 2H medium carbon steel nuts are recommended. Bolts of type A193 B7M have the same chemical composition and heat treatment as bolts of type B7, but with lower mechanical strength and hardness. B7M bolting is commonly used for carbon steel piping and valves in sour service or hydrogen sulfide (H_2S)-containing service. Reducing the material hardness is one of the strategies to prevent sour corrosion caused by H_2S. Consequently, B7M bolts are popular for sour services in the oil and gas industry. The ASTM A194 2HM nuts are compatible with the B7M bolts. Compared to 2H nuts, 2HM nuts have the same chemical composition and heat treatment, but less mechanical strength and hardness. LTCS piping and valves are incompatible with B7 and B7M bolts and 2H and 2HM nuts. Bolt and nut materials proposed for LTCS body valves are A320 L7 bolts and A194 Gr.4 or 7 nuts. Bolts complying with A320 L7 are chromium and molybdenum low-alloy steel bolts similar to bolts complying with A193 B7 and B7M; a Charpy impact test indicates that these bolts are suitable for use at a minimum temperature of –46°C. A194 Gr.4 or 7 nuts are also carbon steel nuts, just like 2H and 2HM bolts; according to a Charpy impact test, they can also be used at a minimum temperature of –46°C. A320 L7 bolts have a size limitation of 2½". Therefore, A320 L43 is a suitable bolt material grade for size over 2½" for LTCS piping and valve body materials. A320 L7M bolts and A194 Gr.4M or 7M nuts are compatible with LTCS piping and valves in sour service or hydrogen sulfide (H_2S)-containing service. For protection against corrosion, these bolts and nuts are coated with molten zinc, also known as hot-dip galvanization (HDG). Table 2.2 summarizes offshore pipeline valves' bolts and nuts materials.

2.2.4 Pup Piece and Transition Piece

A transition piece is a forging item with two different thickening values on each side. If the thickness differential between the valve and pipe (pup piece) exceeds 50%, a transition piece may be welded between the pup piece and the valve's body as a thickness converter. Refer to the ASME B31.8 code, Gas Transmission and Distribution Piping Systems. To compensate for the relatively large thickness difference between the valve and the pipeline, a transition piece is usually welded to the valve's body from one side, and the pup piece from the other side. Typically, the transition piece is made from the same material as the valve's body (e.g., carbon steel or LTCS) to facilitate easier welding. Two materials that are dissimilar may prove difficult to weld. Pup pieces may be created in a different material than transition pieces, and they typically use the same material as the pipelines connected to them. Material engineers upgraded the section of pipeline on the platform from carbon steel to 22Cr duplex even though carbon steel is capable of withstanding well-treated oil and gas fluids without any corrosion risk. By upgrading from carbon steel to 22Cr duplex, the wall thickness of pipelines can be reduced as 22Cr has a higher mechanical and

TABLE 2.2
Bolting Materials for Offshore Pipeline Valves

Offshore Pipeline Valve Body Materials	Bolt Materials	Nut Materials
Carbon steel	A193 B7 + HDG (low-alloy steel)	A194 2H + HDG (carbon steel)
Carbon steel in sour service	A193 B7M + HDG (low-alloy steel)	A194 2HM + HDG (carbon steel)
LTCS	A320 L7 + HDG (low-alloy steel) Bolt size maximum 2½" A320 L43 + HDG (low-alloy steel) Bolt size larger than 2½"	A194 Gr.4 or 7 + HDG (carbon steel)
LTCS in sour service	A320 L7M + HDG (low-alloy steel)	A194 Gr.4M or 7M + HDG (carbon steel)

corrosion resistance than carbon steel. In Figure 1.7, a 38" valve is depicted in LTCS, which will be welded to a 22Cr duplex pipeline in the construction yard. As a result, the transition piece is made of LTCS, and the pup piece is made of 22Cr duplex.

2.2.5 Seal

Leakage can occur either externally to the environment from the valve or internally within the valve. External leakage occurs between the body pieces and between the stem and body. The internal leak is caused by spaces between the valve seats and the body. In offshore pipeline valves, two types of seals are used to prevent leaks: a lip seal and a graphite ring. Lip seals are more resistant to abrasion than elastomeric O-rings; they feature antiexplosive decompression (AED) as a default feature and can handle a wide range of chemicals, temperatures, and pressures. A rapid reduction in the pressure of a gaseous media causes explosive decompression (also known as rapid gas decompression or RGD) of elastomer seals and O-rings. When the pressure is rapidly released, gas that has permeated the elastomer seal expands violently, resulting in fissuring and seal failure. Lip seals are usually composed of a soft (non-metallic) material, such as Teflon, energized by a metallic spring in a CRA, such as Elgiloy® (Figure 2.6). Elgiloy® is a non-magnetic cobalt–chromium–nickel–molybdenum alloy that offers a unique combination of high strength, excellent formability, excellent corrosion resistance, and excellent fatigue strength. Graphite has several advantages, including resistance to extreme low and high temperatures from –200°C to 550°C, corrosion, and fire. Valve designs should be fire safe in order to minimize fugitive emissions from external leaks and internal leaks during a fire. As a secondary seal and backup for lip seals, graphite seals are used to seal the valves' parts in the event of a fire when the lip seals are melted.

FIGURE 2.6 Lip seals with Teflon coatings in white colors and metallic springs. (Photo by the author.)

2.2.6 Stem Bearing

A bearing is a circular component installed around the valve stem to prevent the stem side from moving due to the applied loads. The stem of a pipeline ball valve rotates, causing constant friction with the bearing. Since the stem of a pipeline TCG valve travels linearly up and down, the friction between the stem and the bearings is greater. The metal-to-metal contact between the valve stem and the bearing under load causes fretting corrosion and contributes to wear of these two components. Wear between the valve stem and the bearing increases the force required to operate the valve. The term fretting corrosion refers to the deterioration of material at the interface of two contacting surfaces caused by movement or sliding. There is a Teflon coating on the internal part of the bearing to prevent metal-to-metal contact with the valve stem and fretting corrosion. The metallic parts of the stem bearings are constructed of austenitic SS 316. The thrust washer is another type of bearing that is used to prevent the stem from being thrown out of its position. Thrust washers in offshore pipeline valves are recommended to be constructed from SS 316 with Teflon lining. The austenitic SS s of the 300 series contain approximately 18% chromium and 8% nickel. When nickel is added to SSs, they acquire an austenitic structure, as well as other desirable properties such as formability, ductility, weldability, toughness, and low-temperature properties. The SS 316 can provide the highest level of corrosion resistance among other types of austenitic SSs in chloride and offshore environments.

2.2.7 Spring

In ball valves and through conduit gate valves, springs are utilized for various purposes. First, in the ball and through conduit gate valves, a spring is located behind the valve seat to push the seat toward the closure member (ball or disk) to provide sealing between the ball/disk and seat and prevent internal leakage (Figure 2.7). In addition, springs are used as antistatic devices between the ball/disk and the stem and body of the valve (Figure 2.8) to ensure electrical continuity between the valve closure member, valve body, and the stem. Antistatic springs facilitate discharging the static

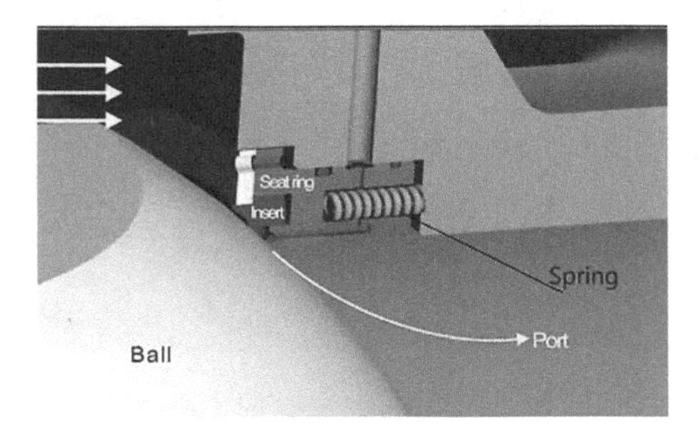

FIGURE 2.7 Ball valve contains a spring behind the seat. (Photo by the author.)

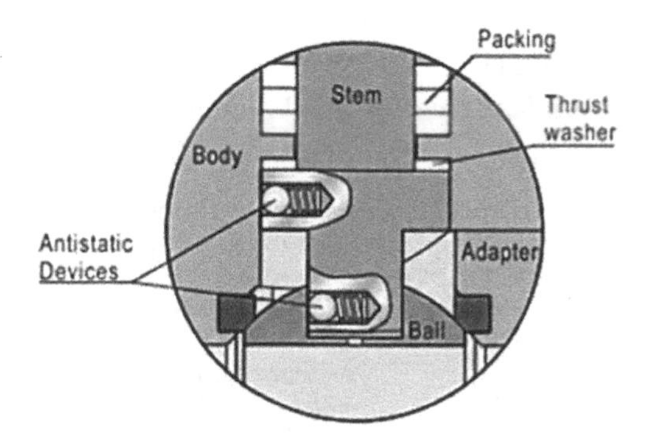

FIGURE 2.8 The antistatic springs of a ball valve.

electricity produced in the valve due to friction between the valve parts during the operation and minimize the risk of fire and exposition of the valves. Inconel X750 or Elgiloy® are two popular spring materials.

According to the cross-section drawing in Figure 1.10, pipeline valve parts are listed in Table 1.1. The objective of this section is to provide selected materials for each of the parts, as indicated in Table 2.3.

TABLE 2.3
List of Parts and Materials of the Pipeline Valve Shown in Figure 1.10

Item Number	Designation	Material
1	Body	LTCS, ASTM A352 LCC + Inconel 625 partially weld overlay
2	Bonnet	LTCS, ASTM A350 LF2 + Inconel 625 partially weld overlay
3	Ball	13Cr-4Ni, ASTM A182 F6NM + tungsten carbide (TC)
4	Stem	13Cr-4Ni, ASTM A182 F6NM
6	Seat ring or seat	13Cr-4Ni, ASTM A182 F6NM + tungsten carbide
7	Cover	SS316, ASTM A182 F316
18	Pup piece	22 Cr duplex, ASTM A182 F51
25	Antistatic ball	SS316
26	Stem key or shaft key	13Cr-4Ni, UNS 415000
34	Vent plug	LTCS, ASTM A350 LF2
36	Drain plug	LTCS, ASTM A350 LF2
41	Injection plug	SS316, ASTM A182 F316
42	Injection plug extension	SS316, ASTM A182 F316
43	Mini check valve	SS316, ASTM A182 F316
300	Body stud bolt	Cr-Mo low-alloy steel, ASTM A320 L7 + hot-dip galvanized (HDG)
301	Body nut	Carbon steel, ASTM A194 Gr.7 + hot-dip galvanized (HDG)
502	Fire safe seal (graphite)	Graphite with corrosion inhibitor
503	O-ring gasket	Viton®*
507	Stem seal (lip seal)	Teflon + Elgiloy® spring

(*Continued*)

TABLE 2.3 (CONTINUED)
List of Parts and Materials of the Pipeline Valve Shown in Figure 1.10

Item Number	Designation	Material
510	Fire safe seal (graphite)	Graphite with corrosion inhibitor
511	Stem seal (lip seal)	Teflon + Elgiloy® spring
521	Seat to body seal (lip seal)	Teflon + Elgiloy® spring
522	Fire safe seal (graphite)	Graphite with corrosion inhibitor
523	Seat to body seal (lip seal)	Teflon + Elgiloy® spring
524	Seat insert	Not applicable for pipeline valves because they are metal seated. 13Cr-4Ni, ASTM A182 F6NM + tungsten carbide
530	Seat scraper	Teflon or an SS316 metallic ring
650	Lower stem bearing	SS316 + Teflon lining
651	Thrust washer	SS316 + Teflon lining
652	Lower ball bearing	SS316 + Teflon lining
653	Thrust washer	SS316 + Teflon lining
655	Upper ball bearing	SS316 + Teflon lining
657	Upper ball bearing	SS316 + Teflon lining
700	Spring	Inconel X750 or Elgiloy®
703	Antistatic spring	SS316

* Viton® is the trade name for a family of fluorocarbon rubber polymeric rubbers commonly used in sealing applications due to their chemical compatibility and temperature resistance. Rubbery materials are elastomers, which are capable of returning to their original shape after being stretched.

2.3 COATING

The pipeline valves' body, bonnet, and transition pieces are in non-CRA carbon steel or LTCS and have contact with the corrosive offshore atmosphere. Carbon or LTCS can be corroded significantly in the offshore environment, requiring a proper coating to prevent external corrosion protection. Depending on the operating temperature of the valve and the presence of insulation surrounding the valve, one could apply an inorganic zinc-rich primer or thermal spray aluminum (TSA) coating. The pipeline valve shown in Figure 2.5 has an operating temperature of less than 120°C, and it has a white-color coating called inorganic zinc-rich primer. TSA is an alternative coating proposed for carbon steel body valves if the operating temperature exceeds 120°C and/or the valve is insulated and be at the risk of corrosion under insulation (CUI). CUI is a severe localized and external type of corrosion that occurs on insulated facilities and components such as industrial valves. The word primer means the first layer of coating applied on the steel. The applied coating creates a layer between the steel and the corrosive atmosphere. In addition, the coating contains zinc that creates ***cathodic protection*** **or** ***galvanic effect***. Zinc, as a sacrificial metal, forms an anode and is corroded in favor of the metal surface in contact with it. Thus, the metal surface in contact with the zinc in the paint is protected.

Non-CRA low-alloy steel bolts and carbon steel nuts do not require zinc-rich primer because they are corrosion-proofed through HDG. HDG is a process for coating steel with a zinc layer by passing the metal through molten zinc. In practice, it has been observed offshore that the HDG is removed from bolts and nuts after some years of operation; the unprotected carbon or low-alloy steel bolts and nuts then rust and corrode. During construction, the valve pieces will be welded to the pipeline. Due to the fact that welding can weaken or damage the coating, it is common to apply a protective coating after welding. The valve pup piece should be partially coated, so that the end that is welded to the pipeline remains uncoated. During the transportation and handling of the pup piece, the uncoated portion of the pup piece can be protected against external corrosion with a primer coating before welding to the pipeline. In the construction yard, the uncoated section of the pup piece is coated after welding between the valve pup piece and the pipeline.

QUESTIONS & ANSWERS

1. What is the correct statement regarding material selection for offshore pipeline valves?

   A. Valve bodies are usually made of Inconel 625.
   B. Pipeline valves can have internals constructed of carbon steel.
   C. The 17-4PH stem is an excellent material choice for stems.
   D. Tungsten carbide hardfacing is used to coat and protect the body and seat contact surfaces.

   Answer) Option A is not valid as the offshore valves are constructed from carbon steel or low-temperature carbon steel. The valves' internals made of carbon steel are heavily corroded, making option B incorrect. Materials such as 13Cr-4Ni are suitable for the internals of offshore valves. Since 17-4PH is not suitable for valve stems in marine environments, option C is incorrect. Therefore, option D is the correct response.

2. The body of a pipeline valve is constructed of carbon steel and is attached to a 22Cr duplex pipeline on the platform. What are the materials used to construct the transition and pup pieces?

   A. 22Cr duplex transition pieces and carbon steel pup pieces
   B. LTCS transition pieces and 22Cr duplex pup pieces
   C. Carbon steel transition pieces and 22Cr duplex pup pieces
   D. 22Cr duplex transition and pup pieces

   Answer) Option C is the correct one since the material of the transition piece is the same as the material of the valve's body, and the material of the pup piece is the same as the material of the pipeline connected to it.

3. Which part of the valve will be coated with hot-dip galvanizing?

   A. Body and bonnet
   B. Pup pieces
   C. Ball and seats
   D. Bolting (bolts and nuts)

   Answer) Option D is the correct answer.

4. In relation to coating offshore pipeline valves for external corrosion protection, what is the correct statement?

    A. The body and bonnet of offshore pipeline valves are always coated.
    B. Coating should extend the entire length of pup pieces.
    C. TSA coating should always be applied.
    D. All other options are incorrect.

    Answer) It is appropriate to choose option A. Pipeline valves are constructed of materials such as carbon steel that are not corrosion resistant. Because of this, they must be protected from external corrosion by coatings. The pup piece is partially coated in order to prevent the coating from being damaged during welding to the pipeline, so option B is incorrect. In fact, option C is incorrect, since a zinc-rich inorganic primer can also be used for coating offshore pipeline valves. Therefore, option D is wrong since option A is correct.

5. Select the valve part that is only made of metal.

    A. Stem bearing
    B. Thrust washer
    C. Stem
    D. Lip seal

    Answer) The stem bearing, thrust washer, and lip seal are made of both metallic and non-metallic pieces. As the stem is the only part comprised of metal, option C is the correct answer.

BIBLIOGRAPHY

1. American Petroleum Institute (API) Recommended Practice (RP) 615. (2010). *Valve selection guide*. 1st edition. Washington, DC: API.
2. American Society of Mechanical Engineers (ASME) B31.3. (2012). *Process piping*. New York, NY: ASME.
3. American Society of Mechanical Engineers (ASME) B31.8. (2020). *Gas transmission and distribution piping systems*. New York, NY: ASME.
4. Miller M.J. (2009). *In-field welding and coating protocols*. USA: Gas Technology Institute.
5. Precision Polymer Engineering (PPE) (2021). *Explosive decompression resistant seals*. [online] Available at: https://www.prepol.com/industries/oil-gas/explosive-decompression [access date: 14th December, 2021].
6. Sanctis M. et al. (2017). Study of 13Cr-4Ni-(Mo) (F6NM) steel grade heat treatment for maximum hardness control in industrial heats. *Metals*, 7(9) 351. https://doi.org/10.3390/met7090351
7. Sotoodeh K. (2018). Pipeline Valves Technology, Material Selection, Welding, and Stress Analysis (A Case Study of a 30 in Class 1500 Pipeline Ball Valve). *American Society of Mechanical Engineers (ASME), Journal of pressure vessel technology*. 140(4): 044001. https://doi.org/10.1115/1.4040139. Paper No. PVT-18-1043.

8. Sotoodeh K. (2019). Valve operability during a fire. *American Society of Mechanical Engineers (ASME), Journal of Offshore Mechanics and Arctic Engineering*, 141(4): 044001. https://doi.org/10.1115/1.4042073. Paper No. OMAE-18-1093.
9. Sotoodeh K. (2020). Optimized material selection for subsea valves to prevent failure and improve reliability. *Journal of Life Cycle Reliability and Safety Engineering*, Springer. https://doi.org/10.1007/s41872-020-00152-x
10. Sotoodeh K. (2021). *A practical guide to piping and valves for the oil and gas industry*, (1st edition). Austin, USA: Elsevier (Gulf Professional Publishing).
11. Sotoodeh K. (2021). Dissimilar welding between piping and valves in the offshore oil and gas industry. *Welding International*, Taylor and Francis. https://doi.org/10.1080/09507116.2021.1919495

3 Double Isolation and Bleed Concept

3.1 INTRODUCTION

A ball valve or through conduit gate (TCG) pipeline valve is used for the start/stop of fluid in a pipeline system. In accordance with API 6D and ISO 14313, standard for pipeline and piping valves as well as pipeline transportation systems-pipeline valves, a double isolation and bleed (DIB) valve is a valve with two seating surfaces, each of which, in the closed position, provides a seal against pressure from a single source, with a means of venting/bleeding the cavity between the seating surfaces. According to the above standards, DIB valves must have at least one bidirectional seat. Double piston effect (DPE) and single piston effect (SPE) seats are synonymous with bidirectional and unidirectional seats, respectively. In addition, SPE seats are also referred to as "***self-relieving (SR)***." Valve manufacturers provide SPE seats as a standard feature and DPE seats as an optional feature. Figure 3.1 illustrates the seat and ball contact in a ball valve, and the blue line indicates fluid flow from left to right. Regardless of whether the seat is SPE or DPE, it is pushed against the ball by both spring force and fluid pressure, preventing internal fluid leakage into the body cavity.

However, DPE and SPE seat functions are different when the media is trapped in the body cavity. Figure 3.2 illustrates a DPE seat that prevents trapped fluid from leaking into the line from the body cavity.

As the fluid is trapped inside the body cavity of the SPE seat, any increase in media pressure that overcomes the spring force behind the seat can push the seat ring away from the ball and discharge the cavity pressure (Figure 3.3). Valve cavity relief pressure in an SR seat should not exceed 133% of the valve's pressure rating.

The design of a DPE seat differs from that of an SPE seat. Figure 3.4 illustrates a ball valve with an SPE (SR) seat on the left side and a DPE seat on the right side. DPE seats have a longer seat retainer or seat ring containing two lip seals that seal the seat to the body. Due to the fact that a lip seal is a unidirectional seal, two lip seals must be used in parallel between the DPE seat and body. Double lip seals provide bidirectional sealing in a DPE seat and isolation from the cavity to the line and from the line to the cavity.

Selecting a DPE seat instead of an SPE one results in a longer seat configuration and longer valve dimensions face-to-face or end-to-end. In addition, a DPE seat has more friction and more substantial contact with the closure member (a ball in this case), which increases the torque required for the valve to operate (a torque is defined as a force that produces rotation and is applicable to quarter-turn valves). An increased force and torque for valve operation may result in a larger and more expensive valve operator (such as an actuator) and stem. In summary, a valve with a DPE seat may be more expensive than one with an SPE seat. A valve with two SPE or SR seats is considered a standard design. An API 6D definition states that a valve with two DPE

DOI: 10.1201/9781003343318-3

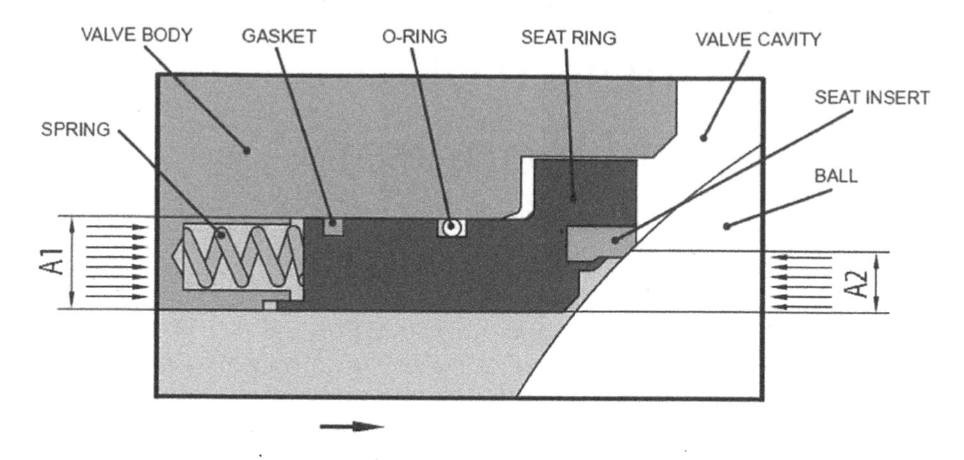

FIGURE 3.1 The ball valve seat closes against the ball. (Photo by the author.)

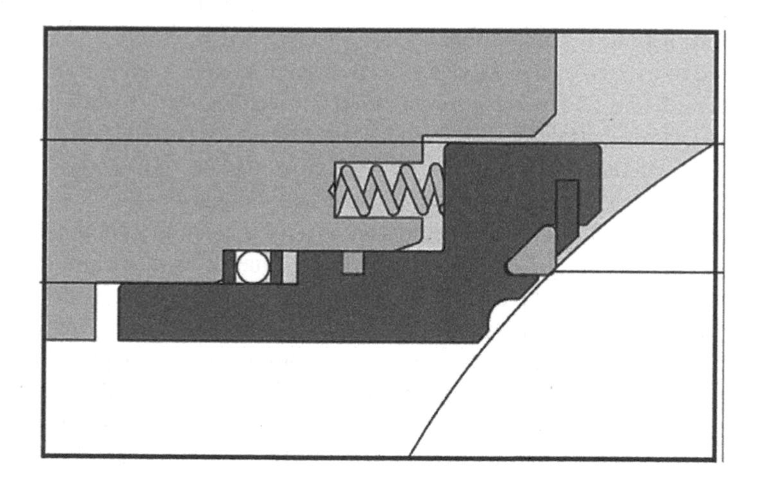

FIGURE 3.2 Trapped fluid in the cavity for the DPE seat. (Photo by the author.)

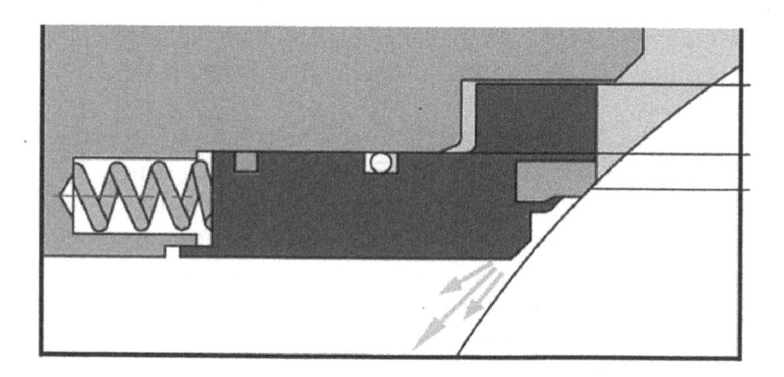

FIGURE 3.3 Fluid discharge from a body cavity to the line in a self-relieving seat. (Photo by the author.)

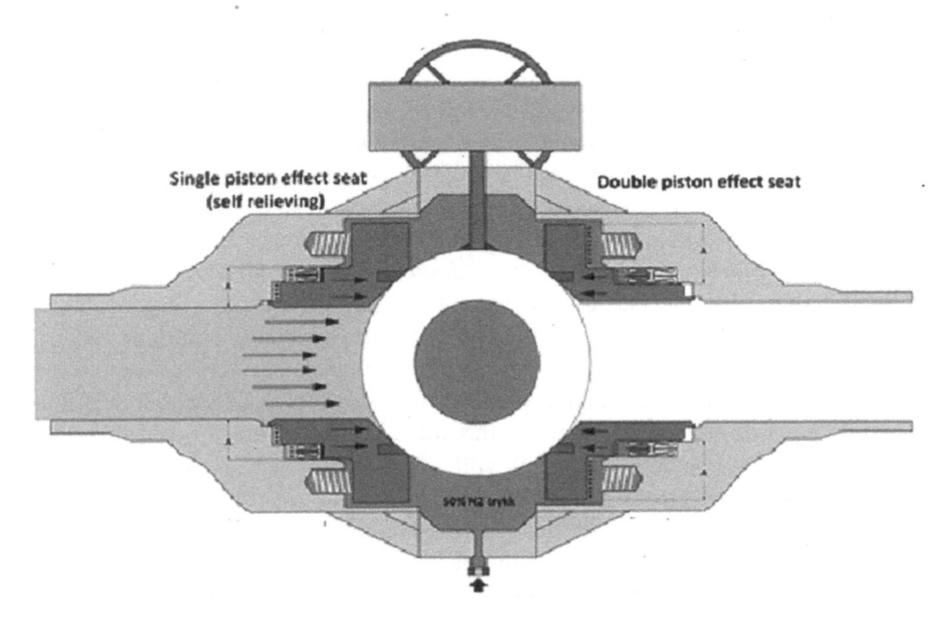

FIGURE 3.4 Configuration of DPE and SPE seats in a ball valve. (Photo by the author.)

seats is called a DIB1 valve, and a valve with one DPE seat and one SPE seat is called a DIB2 valve. In general, floating ball valves, trunnion-mounted ball valves with two SPE seats, and slab gate valves with a fixed seat cannot be used as DIB valves. DIB1 is more expensive and has a more complex design than DIB2. As a result, valve engineers should be aware of the fact that DPE seats in the DIB1 valve cannot release trapped pressure inside the cavity in the line because both seats are DPE. The trapped liquid pressure in the cavity of a DIB1 valve in liquid service can cause excessive force since liquids are not compressible, unlike gases. As a result, a pressure relief valve (PRV) is installed on the DIB1 valve's body cavity to alleviate the added pressure that could result in penetration or leakage. Safety valves, such as the PRV, are used to limit the pressure in a piping system, including industrial valves. In summary, a DIB2 valve has the following advantages over a DIB1 valve:

- Less costly;
- Less complexity in design;
- Less torque/force for operation leads to a possible cheaper valve operator and stem;
- More compact seat design (only applicable to one seat);
- Shorter face-to-face or end-to-end dimensions.

3.2 PIPELINE VALVES DIB CONCEPT

As the most essential and complex valves in an offshore unit, pipeline valves are expected to perform to a superior level of safety and reliability. Providing double isolation or dual barrier arrangements for pipeline valves can improve the safety and reliability of these valves during operation and maintenance by preventing possible fluid leakage. DIB valves are typically selected for use in high-pressure classes of piping (e.g., Class 600, which equals 100 bar or greater nominal pressure) as well as hazardous services, such as flammable hydrocarbons and toxic services (e.g., hydrocarbon oil and gas). Fluids that are toxic might cause permanent injury or even death to operators or maintenance personnel. Pipeline valves are usually found in pipelines carrying high pressures and hydrocarbons that are flammable.

In order to achieve double isolation, a "***double block and bleed (DBB)***" arrangement is employed by two barriers: a couple of standard valves with a bleeder valve to release the trapped fluid between them. Due to the implementation of three valves, a DBB arrangement is costly, bulky, and space consuming. An alternative, more economical method of double isolation would be to use a single valve with a DIB function and a bleed-off connection from the valve cavity. Figure 3.5 illustrates and compares a DBB schematic on the left with a DIB on the right.

As depicted in Figure 3.6, three offshore pipeline valves are located downstream (after) a pig launcher. A pig launcher is a pressure-contained facility used to shoot a pig through a pipeline. An operator opens the first pipeline valve after the pig launcher for shooting the pig inside the pipe and closes it again once the pig is passed through it. The first pipeline valve located after the pig launcher is a normally closed valve designated as "NC," which signifies "normally closed." In addition, a second pipeline valve with the symbol "HV" acts as a backup for the first one and remains open during operation. The first and second valves after the pig launcher both have electrical actuators. Third, the third pipeline valve is a normally open valve with a hydraulic actuator, and its symbol is "SDV," which stands for shutdown valve. After the pig launcher, all three pipeline valves should provide double isolation for safety and reliability reasons, especially during valve maintenance. Both DBB and DIB configurations can provide double isolation. A DBB arrangement requires having six pipeline valves in series rather than three, which is a costly design and therefore avoided. Instead, the valve engineer utilizes the DIB design with one or two DPE seats for every pipeline valve. The previous explanations indicate that DIB2 design offers several advantages over DIB1: lower costs, lower design complexity, reduced torque requirements, etc. In addition, DIB2 design for pipeline valves due to maintenance considerations is considered over-design, which is explained in more detail in the following paragraph.

This section discusses three maintenance scenarios: the first scenario is for the pig launcher or pipeline between the pig launcher and the first valve, the second for the pipeline between the first and second valve, and finally, for the pipeline between the

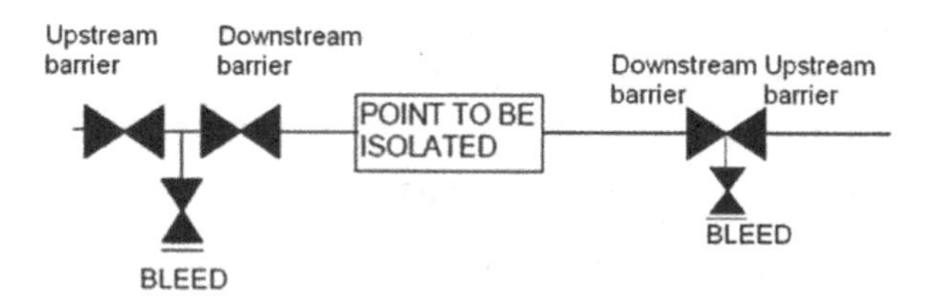

FIGURE 3.5 The DBB schematic on the left and the DIB schematic on the right.

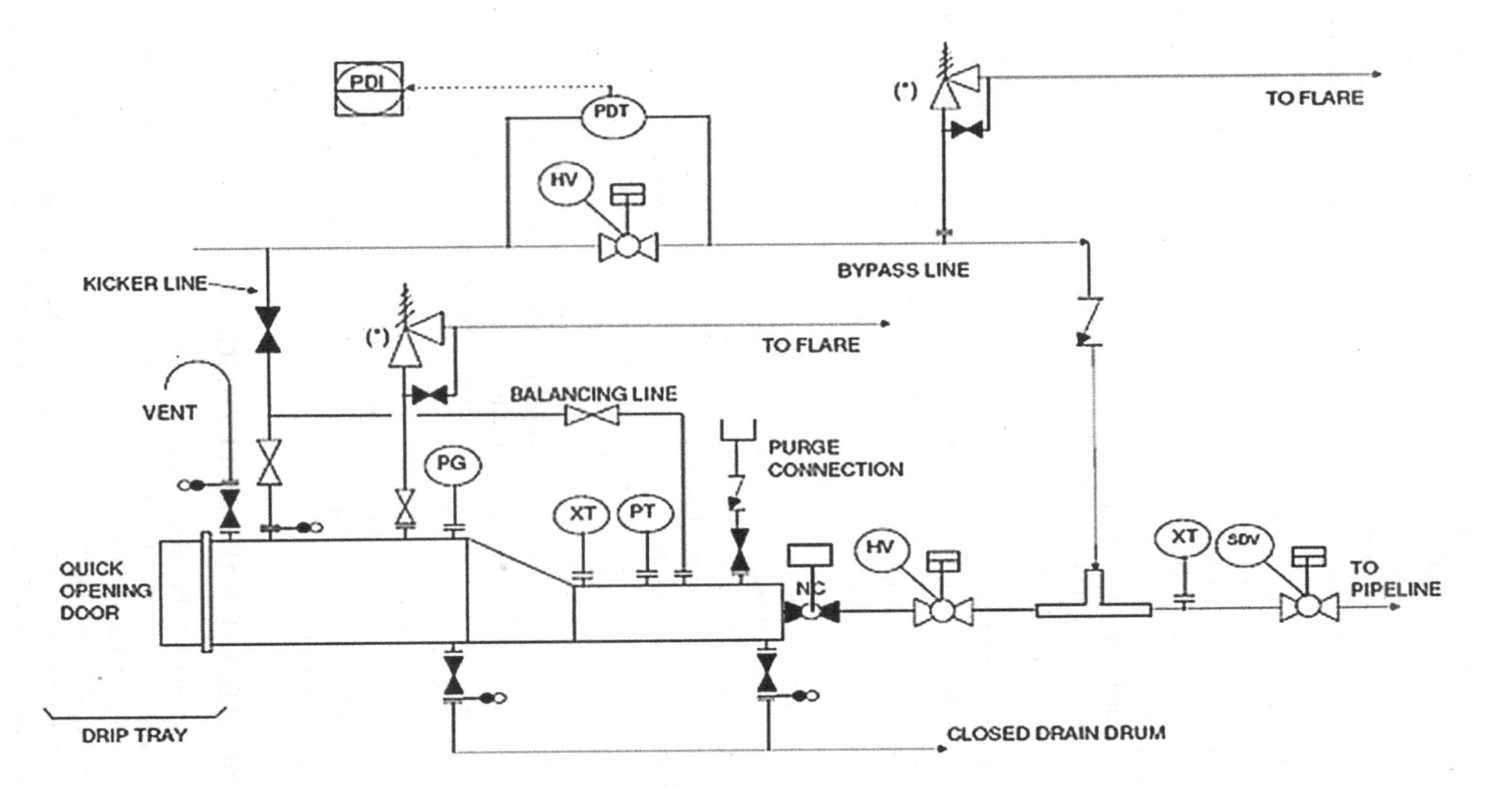

FIGURE 3.6 A pig launcher arrangement including three pipeline valves. (Photo by the author.)

second and third valve. As indicated in the given schematic, a tee is between the second and third pipeline valves, which receive the produced oil from the vertically oriented pipe called oil production line.

Suppose that on-site personnel perform maintenance on the pig launcher. Upon closing the first pipeline valve, the pig launcher and maintenance personnel are isolated from the oil service in the production line. The first pipeline valve may have the DIB2 configuration, in which case the DPE seat may be on the right or left. If the DPE seat is located on the downstream or right side and fails to seal, then the produced oil will enter the valve cavity. In such a case, the SPE or SR seat on the left side of the first pipeline valve is unable to seal against the pressure in the cavity. Therefore, the trapped fluid flows through the SR seat toward the pig launcher and poses a threat to the health and safety of the maintenance personnel. Let us assume that the first pipeline valve is configured in a DIB2 configuration with the DPE seat on the upstream or left side. Therefore, in this instance, the result is the same as the DIB1 selection. This means that if the downstream (right side) SR seat for the DIB2 valve or the downstream DPE seat for the DIB1 valve fails to seal, oil enters the valve cavity. However, the upstream DPE seat in both DIB1 and DIB2 prevents oil leakage toward the pig launcher and maintenance personnel. Taking into account all the benefits of DIB2 over DIB1, the author recommends a DIB2 design with a downstream SR and an upstream DPE seat for the first pipeline valve. Because the DIB2 valve is not a bidirectional valve, it has a preferred installation direction, so the DPE seat should be located on the left (upstream) side of the valve where maintenance occurs.

For the second maintenance condition performed on the pipeline between the first and second valve, a DIB2 configuration with a DPE seat on the upstream or left side is proposed for the second pipeline valve due to the same reasons as for the first pipeline valve. Let us now consider the third maintenance condition between the second and third pipeline valve. The maintenance personnel must be protected against leakage from two directions: one from the oil production line in the schematic and the second from the backflow of the produced oil through the pipeline connected to the right side of the third pipeline valve. If maintenance is to be performed between the second and third pipeline valves, a HV valve on the oil production line should be closed. To protect maintenance personnel, the HV valve on the oil production line must have the DIB2 configuration with a DPE seat on the downstream side closer to the tee. A third pipeline valve should have a DIB2 design with a DPE seat on the left and an SR on the right. Figure 3.7 illustrates that all three pipeline valves have a DIB2 configuration with a DPE seat to the left of the valves and an SR seat to the right.

3.3 DIB EFFECTS ON THE VALVE DESIGN

DIB design first impacts the valve's DPE seat, which is longer (higher) than the standard SPE (SR) seat. By extending a seat, the end-to-end of the DIB pipeline valves can be increased; however, it should be noted that the end-to-end of the pipeline valves is generally based on the manufacturer's and client's specifications. In addition, a finite element analysis (FEA) is typically performed by the valve manufacturers on the seats and the closure member (e.g., ball) of the pipeline valves to confirm

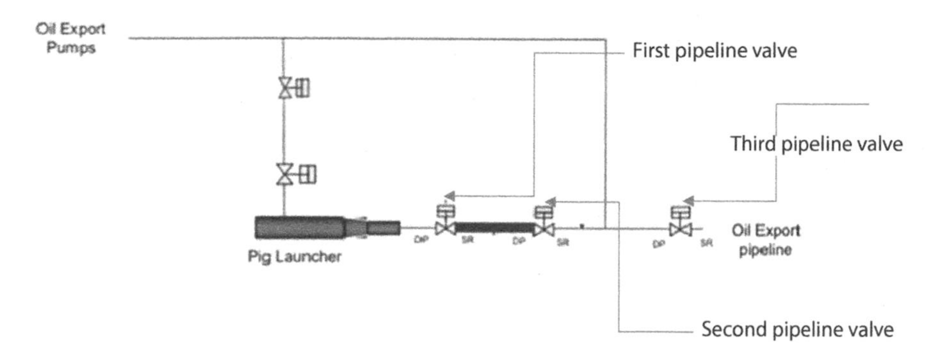

FIGURE 3.7 Three pipeline valve on the oil export line after a pig launcher with a DIB2 design, DPE seats on the left and an SR seat on the right.

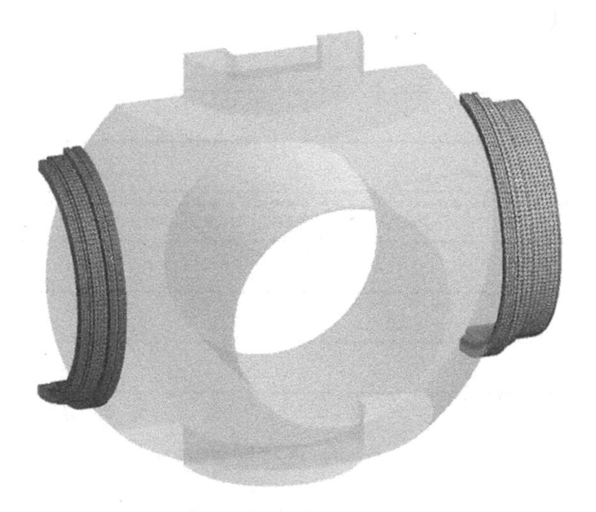

FIGURE 3.8 Modeling of the ball and the seats of a DIB2 pipeline ball valve to apply FEA. (Photo by the author.)

the applied stresses, displacements, and pressure distribution between the components as described above do not damage the valves. Figure 3.8 depicts the modeling of the ball and the seats for a DIB2 pipeline ball valve in order to apply FEA, a computerized method for predicting how a product or component will behave and react under the effects of real-world forces, fluid flow, heat, vibration, and physical effects.

As a valve designer or analyzer, you may take into account different load cases or scenarios, including the design pressure on the SPE seat, the design pressure on the DPE seat, the application of the design pressure on both the DPE and SPE seats simultaneously, and the design pressure within the cavity. It is significant to include the effect of the applied spring's thrust loads to the seat. Figure 3.9 shows the contact pressure between the ball and SPE seat for a 38" CL1500 (pressure nominal of 258 bar) when the SPE seat is pressurized from the line by 1.1 times the nominal pressure. or 284 bar or 28.4 MPa. There are spring thrust loads of 0.14 Mpa and 0.39 Mpa on

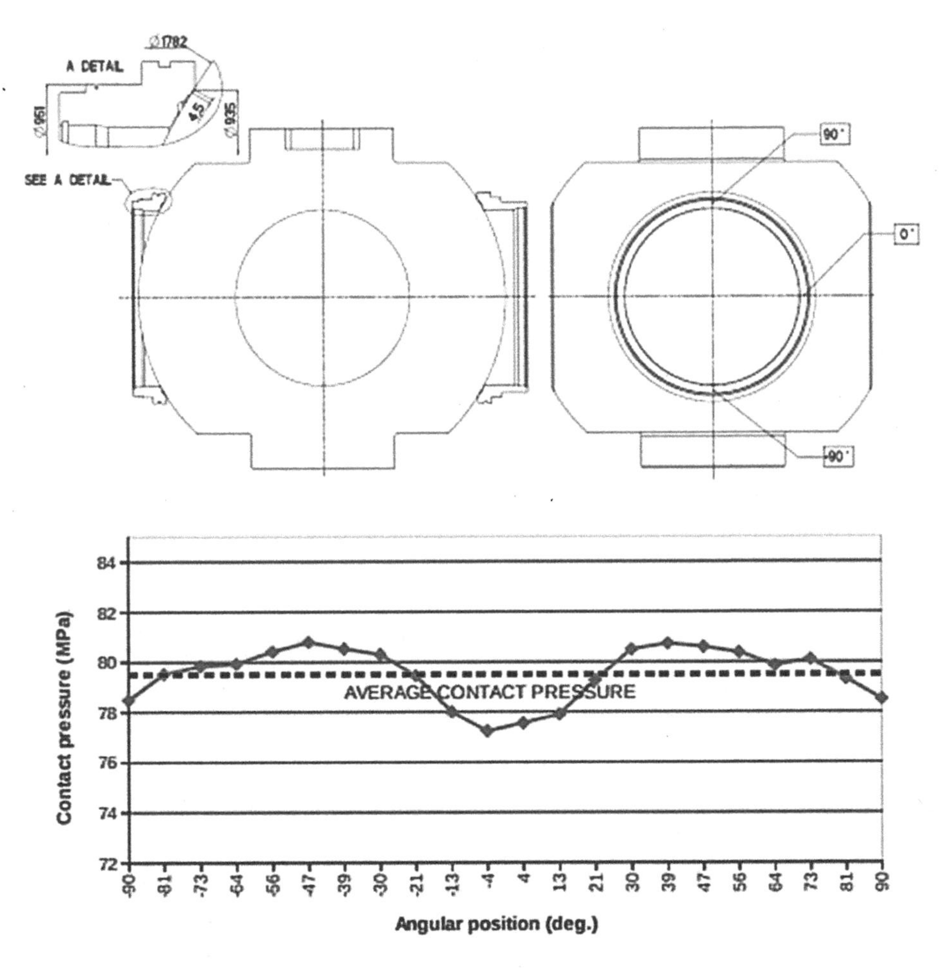

FIGURE 3.9 SPE seat and the ball contact pressure distribution during 284 bar line pressure on the SPE seat.

the SPE and DPE seats, respectively, which are considered in the load case. The spring load on the DPE seat is more than twice that of the SPE seat, indicating that the DPE seat is pushed stronger toward the ball.

Based on the picture, the maximum, average, and minimum contact pressures between the ball and the SPE seat during the SPE seat pressurizing from the line with 284 bar are 80.8, 79.5, and 77.2 Mpa, respectively. Figure 3.10 represents the maximum, average, and minimum DPE seat and ball contact pressure values during the SPE seat pressurizing from a line with 284 bar, respectively: 3, 2.2, and 0.9 MPa.

As illustrated in Figure 3.11, the maximum, average, and minimum DPE seat and ball contact pressure values during the pressurization of the DPE seat from 284 bar are 85.8, 84.1, and 81.8 Mpa, respectively. Because there is a stronger spring applied to the DPE seat, the average contact pressure between the DPE seat and the ball

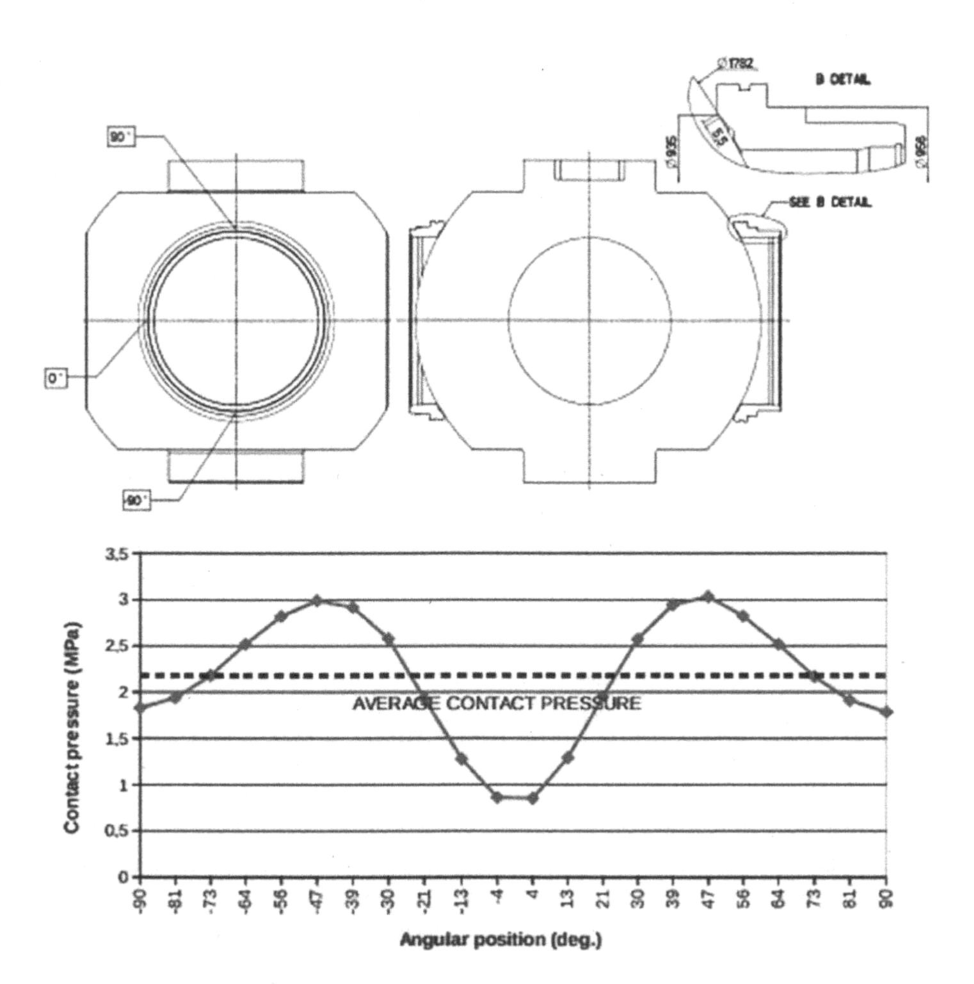

FIGURE 3.10 DPE seat and the ball contact pressure distribution under line pressure of 284 bars on the SPE seat.

during the DPE seat pressurizing from the line with 284 bar is higher than the contact pressure between the SPE seat and the ball during the SPE seat pressurizing from the line with the same pressure. In order to counteract the increased pressure and force between the DPE and the ball, the valve designer may add thickness to the ball which increases the valve weight and its end-to-end dimension. FEA must also be applied to the ball and DPE contact surfaces, in order to ensure that the stress levels between these two components cannot damage them.

The last effect of a DIB configuration on the valve design is the requirement for more torque or force for the valve to operate as compared to a valve with SPE seats due to the higher friction and pressure between the DPE seat and ball. Increasing the force or torque of the valve operation may result in a larger actuator and a stronger stem. The stem can be strengthened by either increasing its diameter or upgrading its

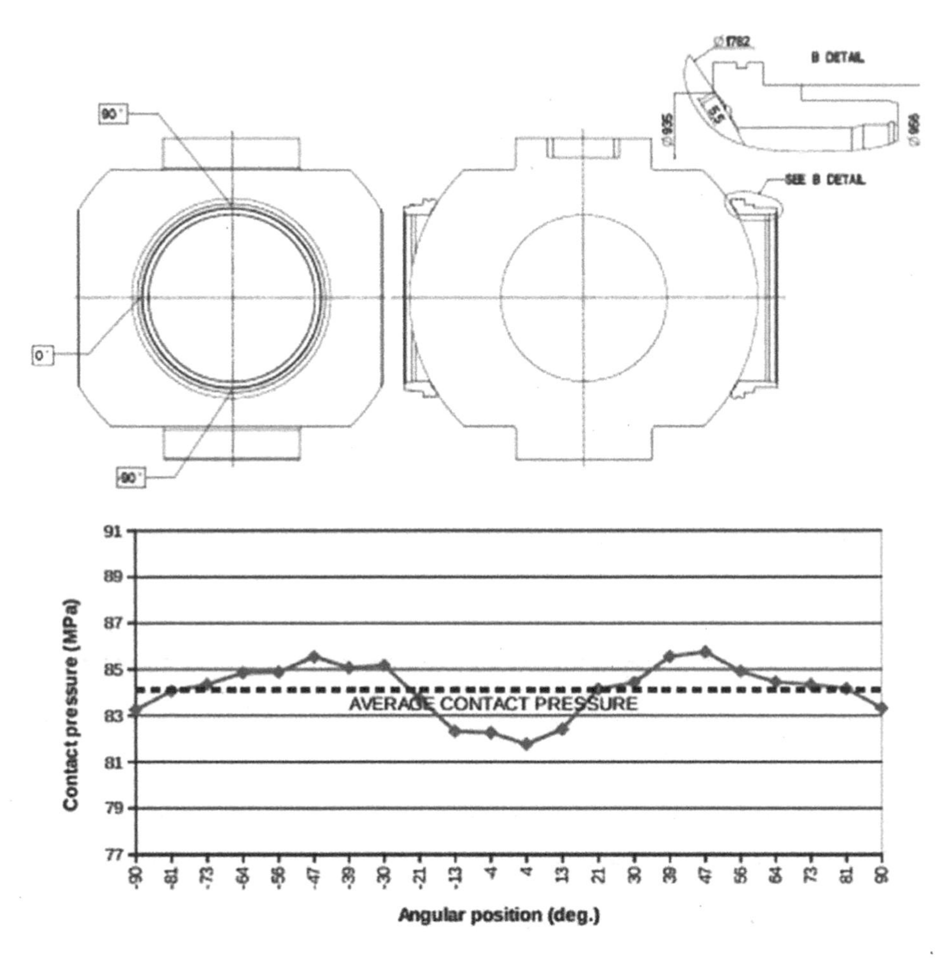

FIGURE 3.11 Pressure distribution on the DPE seat and the ball during a 284 bar line pressure on the DPE seat.

material to one that has a higher mechanical strength. In summary, DIB configuration has the following impact on valve design:

- Longer seat;
- Possibility of a longer end-to-end dimension;
- Possibility of ball enlargement due to an increased contact pressure between the DPE seat and the ball;
- Increasing the required force or torque for the valve to operate;
- Possibly a larger actuator;
- Possibility of enlarging the stem or upgrading the material to a stronger one.

3.4 DIB2 AND VALVE MARKING

As DIB2 valves are not bidirectional, it is essential to specify both SPE and DPE seats' directions in the piping modeling software and in engineering documents such as pipe and instrument diagrams (P&IDs) and isometric drawings. P&IDs are detailed diagrams that are prepared by the process department and show the equipment, piping, and instrumentation. Isometric drawings present a 3D-routed pipe in a 2D plane, highlighting the line number, flow direction, piping components (such as pipes, elbows, tees, types of weld joints), coordinates and elevation of the pipe, and details of the connections between the pipe and equipment. The isometric drawings are one of the most important deliverables of a piping engineering; in most cases, they are used as the main means of communication between piping designers and engineers, on the one hand, and those working on the field and in the construction yard, on the other. On a DIB1 valve identification plate shown in Figure 3.12, the valve manufacturer shall specify the direction of the SPE and DPE seats according to API 6D. There are two arrows on the right indicating the DPE seat, whereas one arrow on the left indicates the SPE seat. This plate should be made of stainless steel 316, with four rivets on each corner made from the same material. Essentially, a rivet is a mechanical fastener with a cylinder, head, and tail that fixes the plate permanently to the valve's body. The identification plate figure proposed in Figure 3.12 is not necessarily used by valve manufacturers. The DPE and SPE, for example, may be engraved on a valve's body by a valve manufacturer, which should be visible, especially after coating the valves. Personnel on site cannot figure out the correct installation direction of DIB valves without the identification plate and without using isometric drawings. When a DIB2 valve is installed incorrectly, it cannot meet the requirements for double isolation.

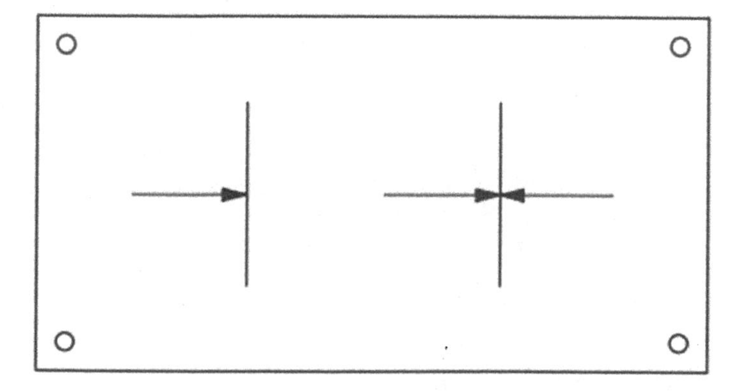

FIGURE 3.12 An example of an identification plate for a DIB2 valve showing one SPE and one DPE seat.

QUESTIONS & ANSWERS

1. Which item does not describe the advantage of DIB2 design over DIB1?

   A. Less expensive
   B. Less design complexity
   C. Higher reliability
   D. Saving a cavity relief valve for liquid services

   Answer) DIB2 cannot always provide greater reliability than DIB1, so option C is the correct response.

2. What is the main advantage of the DIB design as compared to the DBB design?

   A. More compact design
   B. Less costly
   C. Lighter weight
   D. All three other choices are correct

   Answer) Option D is the correct answer.

3. Figure 3.13 illustrates couple of seats for a large size pipeline ball valve. Which sentence is correct about the seats?

FIGURE 3.13 Seats of a pipeline ball valve. (Photo by the author.)

   A. Both seats are SPE or SR and identical.
   B. The seat at the back is DPE, and the one in the front is SPE.
   C. The seat at the back is SPE, and the one in the front is DPE.
   D. Both seats are DPE.

Answer) Two seats are not identical; thus, options A and D are incorrect. In addition, this chapter concludes that the DIB2 configuration with one DPE and one SPE seat is the most optimal design for pipeline valves, provided that the DPE seat is placed on the side of the valve that requires maintenance. A seat at the back is longer (higher) than a seat at the front, so the seat at the back is referred to as a DPE seat, and the one at the front is referred to as an SPE seat. Therefore, option B is correct.

4. Which of the following valve designs requires the installation of a pressure relief valve on the valve's body in order to release the trapped pressure in the cavity?

   A. DIB1 in the liquid service
   B. DIB1 in the gas service
   C. DBB
   D. DIB2

   Answer) Option A is the correct response since DIB1 has two DPE seats which prevent the trapped pressure from being released from the cavity. In addition, unlike gas, liquid is a non-compressible fluid, so it can expand and create excessive pressure inside the cavity. As the gas is compressible, it is not necessary to install a pressure relief valve on the body of the DIB1 valve in the gas service; therefore, option B is incorrect. DBB valves are standard valves with two SR seats capable of releasing excessive pressure in the cavity, so option C is wrong. DIB2 valves have a single SR seat for releasing cavity pressure, so option D is incorrect.

5. Choose the incorrect sentence regarding three pipeline valves after the pig launcher in Figure 3.7.

   A. After the pig launcher, the first valve usually closes and opens during the pig run.
   B. The DPE seats for all three valves are located toward the pig launcher.
   C. The third valve is normally open and is closed in the event of an emergency.
   D. All three valves have an emergency shutdown (ESD) function.

   Answer) Option D is incorrect, as the first two valves after the pig launcher do not have emergency shutdown (ESD) capability. The ESD valve is a highly reliable, actuated valve equipped with a high level of safety and reliability. It is designed to shut down the pipe immediately upon detecting a potentially dangerous event. Process shut down (PSD) valves are similar to ESD valves in that they are actuated to shut off the pipe and fluid flow at a lower level of severity. More information about PSD and ESD valves is given in Chapter 13.

6. What is the proper method of marking DIB2 valves?

   A. Using a tag plate made of carbon steel material
   B. Engraving the valve body in a low-visibility manner
   C. Using a tag plate that only shows the direction of a DPE seat
   D. A tag plate in stainless steel 316 with four rivets in the same material as the tag plate and two sign posts of SPE and DPE on either side

Answer) Option A is incomplete and incorrect since carbon steel tag plates will corrode easily when used in an offshore environment. As a result, option B is incorrect because the engraved marking on the valve should be visible even after coating. Option C is incorrect since the valve manufacturer should indicate both SPE and DPE seats on the tag plate. The correct answer is D.

7. Figure 3.14 illustrates an arrangement for automatically releasing the overpressure from the valve cavity. Which of the following statements about this figure is incorrect?

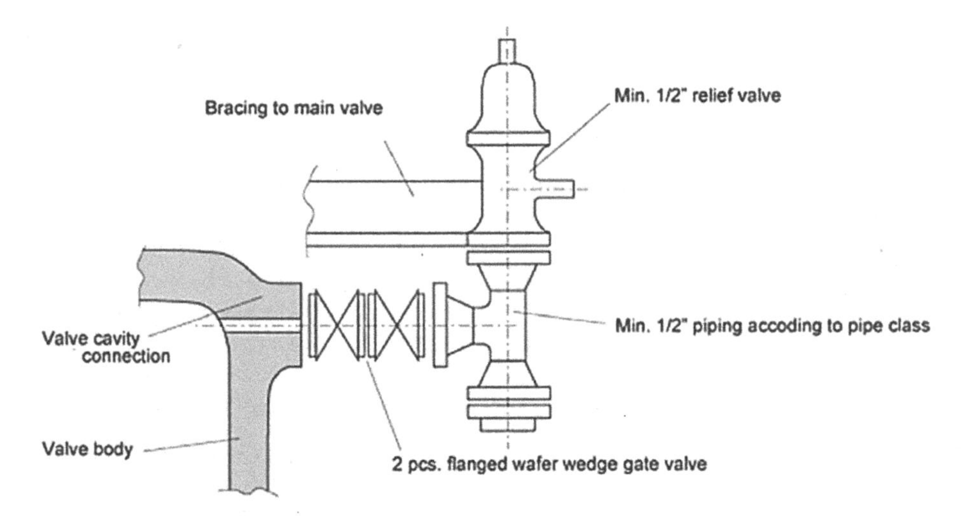

FIGURE 3.14 A cavity relief arrangement. (Photo by the author.)

A. The cavity relief is required for a DIB1 type valve in the gas service.
B. Two wedge gate valves are normally closed, but they will be open during relief valve maintenance.
C. The relief valve is the principal component for releasing trapped and overpressured fluid.
D. Bracing of the relief valve to the valve body is intended to control vibration and absorb shock loads caused by pressure relief valves.

Answer) Option A is incorrect because automatic cavity relief is employed for liquid DIB1 valves, not gas services. Option B is incorrect since two gate valves are normally open and are closed during relief valve maintenance. This is done in order to isolate the valve cavity and trapped fluid from the environment. Options C and D are both correct. To add additional safety through redundancy, two gate valves are designed and operated instead of one. Double isolation is typical in situations where the fluid has a high pressure (e.g., CL600 and higher equal to a nominal pressure of 100 bar) and/or the fluid is toxic or flammable.

BIBLIOGRAPHY

1. American Petroleum Institute (API) 6D. (2014). *Specification for pipeline and piping valves* (24th edition). Washington, DC: API.
2. International Organization for Organization (ISO). (2007). *Pipeline transportation systems—Pipeline valves.* (ISO 14313. 2nd Edition). Geneva, Switzerland: ISO.
3. Raj V. (2021). What is piping isometric drawing? How to read piping isometric drawing? [online] available at: https://www.allaboutpiping.com/how-to-read-piping-isometric-drawing/ [access date: 20th December 20, 2021].
4. Sotoodeh K. (2015). Top entry export line valves design considerations. *Valve World Magazine*, 20(05), 55–61.
5. Sotoodeh K. (2015). Double isolation and bleed ball valves from a design and maintenance point of view. *Valve World Magazine*, 20(11), 44–48.
6. Sotoodeh K. (2018). Valve failures, analysis and solutions. *Valve World Magazine*, 23(11), 48–52.

4 Design for Weight Reduction

4.1 INTRODUCTION

Many oil companies are always looking for ways to develop offshore fields more economically. The pressure to reduce the cost is more prevalent in marginal oilfields that may not generate sufficient earnings, particularly during low oil prices. During the late 1970s and afterward, weight control and reduction gained significant attention in the offshore oil and gas industry. There have been several publications addressing the cost advantages and gains associated with using weight reduction design techniques in the offshore oil and gas industry. After the 1970s, offshore platforms, such as those shown in Figure 4.1, played a key role in the development of oil fields. Oil and gas platforms were initially large and heavy, but as oil and gas production moved deeper into the ocean, operator companies required more compact and economical platforms. As a result, there is limited space and load capacity on offshore platforms. One study has determined that one ton of topside equipment and facilities requires two tons of platform weight for support. As a result, reducing one ton of topside equipment results in a reduction of three tones of platform and topside equipment. Taking into account that topside facilities will cost $20,000 per ton, whereas jacket platforms will cost $13,000 per ton, the savings resulting from a one-ton reduction in topside equipment weight is $46,000.

FIGURE 4.1 An offshore platform. (Courtesy: Shutterstock.)

DOI: 10.1201/9781003343318-4

From a large-scale perspective, the weights of topside facilities depend on the size and number of equipment, layout, pipes, valves, and structures. A reduction in the weight of topside facilities, such as industrial valves, saves costs and facilitates production, lifting, and transportation. A pipeline valve for an offshore pipeline could weigh as much as 70 tons and be challenging to handle and transport.

4.2 DESIGN FOR WEIGHT REDUCTION

The purpose of this chapter is to discuss a number of approaches for reducing the weight of offshore pipeline valves. First, one should calculate the wall thickness and second, one should use circular nuts instead of hexagonal nuts for the body and bonnet connection.

4.2.1 Wall Thickness Calculation and Validation

4.2.1.1 Wall Thickness Calculation

The American Society of Mechanical Engineers (ASME) B16.34, the standard for flanged, threaded, and welding ends covers many aspects of valves, such as pressure–temperature ratings, dimensions, tolerances, materials, and non-destructive examination requirements, testing, and marking. The valve body and bonnet thickness should meet the requirements set in ASME B16.34. The minimum valve body thickness according to ASME B16.34, parameter t_m, is provided in both millimeters and inches. The minimum valve body thickness in ASME B16.34 depends on two valve parameters: internal diameter (parameter *d*) and the pressure class of the valve. Generally, the minimum valve wall thickness increases with the internal diameter and pressure class of the valve. The pressure classes, which are covered by ASME B16.34, are 150 (PN20), 300 (PN50), 600 (PN100), 900 (PN150), 1500 (PN250), 2500 (PN420), and 4500(PN720). "PN" stands for pressure nominal. The internal diameter of a valve is the minimum diameter of fluid passage through the valve bore. The internal diameter of a valve also increases with the size of the valve. The minimum wall thickness according to ASME B16.34 is independent of the type of material used for the valve body. For example, the wall thickness of carbon and stainless steel valve is equal according to ASME B16.34, providing the same internal diameter and pressure class. Table 4.1, extracted from ASME B16.34, provides valve body thicknesses, parameter t_m, based on the valve's internal diameter (parameter *d*) and pressure. For example, a 30" pipeline ball valve in carbon steel body material and CL1500 pressure class equal to 258.6 bar has 625.3 mm internal diameter. Table 4.2 is used to determine internal dimeters of the valves based on size and pressure class. An internal diameter of 625.3 mm would fall between 620 mm and 630 mm. The minimum wall thickness values associated with 620 mm and 630 mm internal diameters with a pressure class of 1500 are 116.9 mm and 118.7 mm, respectively. The values associated with an internal diameter of 625.3 mm can be found using interpolation; in this case, it is 117.8 mm. Thus, a minimum thickness of 117.86 mm is necessary for a valve with an internal diameter of 671.26 mm and pressure class of 1500.

TABLE 4.1
Valve Minimum Body Thickness in Millimeters, Parameter t_m, According to ASME B16.34

Inside Diameter (d) mm	CL150	CL300	CL600	CL900	CL1500	CL2500	CL4500
3	2.5	2.5	2.8	2.8	3.1	3.6	4.9
6	2.7	2.7	3.0	3.1	3.5	4.2	6.5
9	2.8	2.9	3.2	3.4	3.8	4.9	8.0
12	2.9	3.0	3.4	3.7	4.2	5.6	9.6
15	3.1	3.3	3.6	4.2	4.8	6.6	12.0
18	3.3	3.5	3.9	4.7	5.3	7.7	14.3
21	3.5	3.7	4.2	5.2	5.9	8.7	16.7
24	3.7	4.0	4.4	5.7	6.4	9.7	19.0
27	3.9	4.3	4.8	6.3	7.2	11.1	22.2
31	4.3	4.7	5.1	6.6	8.1	12.8	26.1
35	4.6	5.1	5.4	6.9	9.0	14.5	30.0
40	4.9	5.5	5.7	7.2	9.9	16.2	33.9
45	5.2	5.9	6.0	7.5	10.8	17.9	37.9
50	5.5	6.3	6.3	7.8	11.8	19.6	41.8
55	5.6	6.5	6.3	8.3	12.7	21.3	45.7
60	5.7	6.6	6.6	8.8	13.6	23.0	49.6
65	5.8	6.8	6.9	9.3	14.5	24.7	53.6
70	5.9	6.9	7.3	9.9	15.5	26.4	57.5
75	6.0	7.1	7.6	10.4	16.4	28.1	61.4
80	6.1	7.2	8.0	10.9	17.3	29.8	65.3
85	6.2	7.4	8.3	11.4	18.2	31.5	69.3
90	6.3	7.5	8.6	11.9	19.1	33.2	73.2
95	6.4	7.7	9.0	12.5	20.1	34.9	77.1
100	6.5	7.8	9.3	13.0	21.0	36.6	81.0
110	6.5	8.0	10.0	14.0	22.8	40.0	88.9
120	6.7	8.3	10.7	15.1	24.7	43.4	96.7
130	6.8	8.7	11.4	16.1	26.5	46.9	104.6
140	7.0	9.0	12.0	17.2	28.4	50.3	112.4
150	7.1	9.3	12.7	18.2	30.2	53.7	120.3
160	7.3	9.7	13.4	19.3	32.0	57.1	128.1
170	7.5	10.0	14.1	20.3	33.9	60.5	136.0
180	7.6	10.3	14.7	21.3	35.7	63.9	143.8
190	7.8	10.7	15.4	22.4	37.6	67.3	151.7
200	8.0	11.0	16.1	23.4	39.4	70.7	159.5
210	8.1	11.3	16.8	24.5	41.3	74.1	167.4
220	8.3	11.7	17.4	25.5	43.1	77.5	175.2
230	8.4	12.0	18.1	26.6	45.0	80.9	183.1
240	8.6	12.3	18.8	27.6	46.8	84.4	190.9
250	8.8	12.7	19.5	28.7	48.6	87.8	198.8
260	8.9	13.0	20.2	29.7	50.5	91.2	206.6
270	9.1	13.3	20.8	30.8	52.3	94.6	214.5
280	9.3	13.7	21.5	31.8	54.2	98.0	222.3
290	9.4	14.0	22.2	32.8	56.0	101.4	230.2
300	9.6	14.3	22.9	33.9	57.9	104.8	238.0
310	9.8	14.7	23.5	34.9	59.7	108.2	245.9
320	9.9	15.0	24.2	36.0	61.6	111.6	253.7

(*Continued*)

TABLE 4.1 (CONTINUED)
Valve Minimum Body Thickness in Millimeters, Parameter t_m, According to ASME B16.34

Inside Diameter (d) mm	CL150	CL300	CL600	CL900	CL1500	CL2500	CL4500
330	10.1	15.3	24.9	37.0	63.4	115.0	261.6
340	10.2	15.7	25.6	38.1	65.2	118.4	269.4
350	10.4	16.0	26.3	39.1	67.1	121.9	277.2
360	10.6	16.3	26.9	40.2	68.9	125.3	285.1
370	10.7	16.7	27.6	41.2	70.8	128.7	292.9
380	10.9	17.0	28.3	42.2	72.6	132.1	300.8
390	11.1	17.3	29.0	43.3	74.5	135.5	308.6
400	11.2	17.7	29.6	44.3	76.3	138.9	316.5
410	11.4	18.0	30.3	45.4	78.2	142.3	324.3
420	11.5	18.3	31.0	46.4	80.0	145.7	332.2
430	11.7	18.7	31.7	47.5	81.8	149.1	340.0
440	11.9	19.0	32.4	48.5	83.7	152.5	347.9
450	12.0	19.4	33.0	49.6	85.5	155.9	355.7
460	12.2	19.7	33.7	50.6	87.4	159.4	363.6
470	12.4	20.0	34.4	51.7	89.2	162.8	371.4
480	12.5	20.4	35.1	52.1	91.1	166.2	379.3
490	12.7	20.7	35.7	53.7	92.9	169.6	387.1
500	12.9	21.0	36.4	54.8	94.8	173.0	395.0
510	13.0	21.4	37.1	55.8	96.6	176.4	402.8
520	13.2	21.7	37.8	56.9	98.4	179.8	410.7
530	13.3	22.0	38.5	57.9	100.3	183.2	418.5
540	13.5	22.4	39.1	59.0	102.1	186.6	426.4
550	13.7	22.7	39.8	60.0	104.0	190.0	434.2
560	13.8	23.0	40.5	61.1	105.8	193.4	442.1
570	14.0	23.4	41.2	62.1	107.7	196.9	449.9
580	14.2	23.7	41.8	63.1	109.5	200.3	457.8
590	14.3	24.0	42.5	64.2	111.4	203.7	465.6
600	14.5	24.4	43.2	65.2	113.2	207.1	473.5
610	14.6	24.7	43.9	66.3	115.0	210.5	481.3
620	14.8	25.0	44.6	67.3	116.9	213.9	489.2
630	15.0	25.4	45.2	68.4	118.7	217.3	497.0
640	15.1	25.7	45.9	69.4	120.6	220.7	504.9
650	15.3	26.0	46.6	70.5	122.4	224.1	512.7
660	15.5	26.4	47.3	71.5	124.3	227.5	520.6
670	15.6	26.7	47.9	72.5	126.1	230.9	528.4
680	15.8	27.0	48.6	73.6	128.0	234.4	536.3
690	15.9	27.4	49.3	74.6	129.8	237.8	544.1
700	16.1	27.7	50.0	75.7	131.6	241.2	552.0
710	16.3	28.0	50.7	76.1	133.5	244.6	559.8
720	16.4	28.4	51.3	77.8	135.3	248.0	567.7
730	16.6	28.7	52.0	78.8	137.2	251.4	575.5
740	16.8	29.0	52.7	79.9	139.0	254.8	583.4
750	16.9	29.4	53.4	80.9	140.9	258.2	591.2
760	17.1	29.7	54.0	82.0	142.7	261.6	599.0
770	17.3	30.0	54.7	83.0	144.6	265.0	606.9
780	17.4	30.4	55.4	84.0	146.4	268.4	614.7

(Continued)

TABLE 4.1 (CONTINUED)
Valve Minimum Body Thickness in Millimeters, Parameter t_m, According to ASME B16.34

Inside Diameter (d) mm	CL150	CL300	CL600	CL900	CL1500	CL2500	CL4500
790	17.6	30.7	56.1	85.1	148.2	271.9	622.6
800	17.7	31.0	56.8	86.1	150.1	275.3	630.4
820	18.1	31.7	58.1	88.2	153.8	282.1	646.1
840	18.4	32.4	59.5	90.3	157.5	288.9	661.8
860	18.7	33.0	60.8	92.4	161.1	295.7	677.5
880	19.0	33.7	62.2	94.5	164.8	302.5	693.2
900	19.4	34.4	63.5	96.6	168.5	309.4	708.9
920	19.7	35.0	64.9	98.7	172.2	316.2	724.6
940	20.0	35.7	66.2	100.8	175.9	323.0	740.3
960	20.3	36.4	67.6	102.9	179.6	329.6	756.0
980	20.7	37.1	69.0	104.9	183.3	336.6	771.7
1000	21.0	37.7	70.3	107.0	187.0	343.5	787.4
1020	21.3	38.4	71.7	109.1	190.7	350.3	803.1
1040	21.7	39.1	73.0	111.2	194.3	357.1	818.8
1060	22.0	39.7	74.4	113.3	198.0	363.9	834.5
1080	22.3	40.4	75.7	115.4	201.7	370.7	850.2
1100	22.6	41.1	77.1	117.5	205.4	377.5	865.9
1120	23.0	41.7	78.4	119.6	209.1	384.4	881.6
1140	23.3	42.4	79.8	121.7	212.8	391.2	897.3
1160	23.6	43.1	81.2	123.7	216.5	398.0	913.0
1180	23.9	43.7	82.5	125.8	220.2	404.8	928.7
1200	24.3	44.4	83.9	127.9	223.9	411.6	944.4
1220	24.6	45.1	85.2	130.0	227.5	418.5	960.1
1240	24.9	45.7	86.6	132.1	231.2	425.3	975.8
1260	25.2	46.4	87.9	134.2	234.9	432.1	991.5
1280	25.6	47.1	89.3	136.3	238.6	438.9	1007.2
1300	25.9	47.7	90.6	138.4	242.3	445.7	1022.9

TABLE 4.2
Valve Inside Diameter in Millimeters and Inches, According to Valve Nominal Pipe Size (NPS) and Pressure Class, as per ASME B16.34

	CL150		CL300		CL600		CL900		
NPS	mm	inch	mm	inch	mm	inch	mm	inch	DN
1/2"	12.7	0.5	12.7	0.5	12.7	0.5	12.7	0.50	15
3/4"	19.1	0.75	19.1	0.75	19.1	0.75	17.5	0.69	20
1"	25.4	1.00	25.4	1.00	25.4	1.00	22.1	0.87	25
1 1/4"	31.8	1.25	31.8	1.25	31.8	1.25	28.4	1.12	32
1 1/2"	38.1	1.50	38.1	1.50	38.1	1.50	34.8	1.37	40
2"	50.8	2.00	50.8	2.00	50.8	2.00	47.5	1.87	50
2 ½"	63.5	2.50	63.5	2.50	63.5	2.50	57.2	2.25	65
3"	76.2	3.00	76.2	3.00	76.2	3.00	72.9	2.87	80

(*Continued*)

TABLE 4.2 (CONTINUED)
Valve Inside Diameter in Millimeters and Inches, According to Valve Nominal Pipe Size (NPS) and Pressure Class, as per ASME B16.34

	CL150		CL300		CL600		CL900		
NPS	mm	inch	mm	inch	mm	inch	mm	inch	DN
4"	101.6	4.00	101.6	4.00	101.6	4.00	98.3	3.87	100
6"	152.4	6.00	152.4	6.00	152.4	6.00	146.1	5.75	150
8"	203.2	8.00	203.2	8.00	199.9	7.87	190.5	7.50	200
10"	254.0	10.00	254.0	10.00	247.7	9.75	238.0	9.37	250
12"	304.8	12.00	304.8	12.00	298.5	11.75	282.4	11.12	300
14"	336.6	13.25	336.6	13.25	326.9	12.87	311.2	12.25	350
16"	387.4	15.25	387.4	15.25	374.7	14.75	355.6	14	400
18"	438.2	17.25	431.8	17.00	419.1	16.50	400.1	15.75	450
20"	489.0	19.25	482.6	19.00	463.6	18.25	444.5	17.50	500
22"	539.8	21.25	533.4	21.00	511.0	20.12	489.0	19.25	550
24"	590.6	23.25	584.2	23.00	558.8	22.00	533.4	21.00	600
26"	641.4	25.25	635.0	25.00	603.3	23.75	577.9	22.75	650
28"	692.2	27.25	685.8	27.00	647.7	25.50	622.3	24.50	700
30"	743.0	29.25	736.6	29.00	695.2	27.37	666.8	26.25	750
32"	793.7	31.25	787.4	31.00	736.6	29.00	711.2	28.00	-
34"	844.5	33.25	838.2	33.00	781.0	30.75	755.6	29.75	-
36"	895.3	35.25	889.0	35.00	825.5	32.62	800.1	31.50	-
38"	946.1	37.25	939.8	37.00	872.9	34.37	844.5	33.25	-
40"	996.9	39.25	990.6	39.00	920.7	36.25	889.0	35.00	-
42"	1047.7	41.25	1041.4	41.00	965.2	38.00	933.4	36.75	-
44"	1098.5	43.25	1092.2	43.00	1012.6	39.87	977.9	38.50	-
46"	1149.3	45.25	1143.0	45.00	1057.1	41.62	1022.3	40.25	-
48"	1200.1	47.25	1193.8	47.00	1104.9	43.50	1066.8	42.00	-
50"	1250.9	49.25	1244.6	49.00	1149.3	45.25	1111.2	43.75	-
52"	1301.7	51.25	1295.4	51.00	1193.8	47.00	-	-	-
54"	1352.5	53.25	1346.2	53.00	1241.2	48.87	-	-	-
56"	1403.3	55.25	1397.0	55.00	1285.7	50.62	-	-	-
58"	1454.1	57.25	1447.8	57.00	1330.1	52.37	-	-	-
60"	1504.9	59.25	1498.6	59.00	1374.6	54.12	-	-	-

	CL1500		CL2500		
	mm	inch	mm	inch	DN
1/2"	12.7	0.50	11.2	0.44	15
3/4"	17.5	0.69	14.2	0.56	20
1"	22.1	0.87	19.1	0.75	25
1 1/4"	28.4	1.12	25.4	1.00	32
1 1/2"	34.8	1.37	28.4	1.12	40
2"	47.5	1.87	38.1	1.50	50
2 ½"	57.2	2.25	47.5	1.87	65
3"	69.9	2.75	57.2	2.25	80
4"	91.9	3.62	72.9	2.87	100
6"	136.4	5.37	111.0	4.37	150
8"	177.8	7.00	146.1	5.75	200
10"	222.3	8.75	184.2	7.25	250
12"	263.4	10.37	218.9	8.62	300

(*Continued*)

TABLE 4.2 (CONTINUED)
Valve Inside Diameter in Millimeters and Inches, According to Valve Nominal Pipe Size (NPS) and Pressure Class, as per ASME B16.34

	CL150		CL300		CL600		CL900		
NPS	mm	inch	mm	inch	mm	inch	mm	inch	DN
14"	288.8	11.37	241.3	9.50					350
16"	330.2	13.00	276.1	10.87					400
18"	371.3	14.62	311.2	12.25					450
20"	415.8	16.37	342.9	13.50					500
22"	457.2	18.00	377.7	14.87					550
24"	498.3	19.62	412.8	16.25					600
26"	539.8	21.25	447.5	17.62					650
28"	584.2	23.00	482.6	19.00					700
30"	625.3	24.62	517.4	20.37					750
32"	-	-	-	-					-
34"	-	-	-	-					-
36"	-	-	-	-					-
38"	-	-	-	-					-
40"	-	-	-	-					-
42"	-	-	-	-					-
44"	-	-	-	-					-
46"	-	-	-	-					-
48"	-	-	-	-					-
50"	-	-	-	-					-
52"	-	-	-	-					-
54"	-	-	-	-					-
56"	-	-	-	-					-
58"	-	-	-	-					-
60"	-	-	-	-					-

ASME B16.34 mandatory Appendix VI provides an alternative method based on basic equations for obtaining the minimum wall thickness of the valves according to the internal diameter and pressure class of the valve. The minimum wall thickness values provided in Table 4.1 are calculated according to the basic equations provided in mandatory Appendix VI. Table 4.3 summarizes the basic equations for minimum wall thickness calculation as per ASME B16.34, mandatory Appendix VI. In this case, parameter p_c is equal to 1500, and internal diameter or parameter d is between 3 mm and 1300 mm, so $t_m(1500) = 0.18443*d + 2.54 = 0.18443*625.3 + 2.54 = 117.86$ mm.

Tables 4.1 and 4.3 provide very conservative values for wall thickness. This indicates that the wall thickness values provided are relatively high, resulting in heavier and bulkier valves. As a result, ASME B16.34 provides another method for calculating valve wall thickness using Equation 4.1. The minimum wall thickness values, t_m, shown in Table 4.1 or 4.3, calculated based on the equations in mandatory Appendix VI are all greater than the thickness values determined by Equation 4.1.

TABLE 4.3

Basic Equations for Minimum Valve Wall Thickness Calculation, as per ASME B16.34, Mandatory Appendix VI

Valve Pressure Class, P_c	Inside Diameter (d), mm	Metric Equation t_m (mm)	Round
150	$3 \leq d < 50$	$t_m(150) = 0.064d + 2.34$	Off, one decimal
150	$50 \leq d \leq 100$	$t_m(150) = 0.020d + 4.5$	Off, one decimal
150	$100 < d \leq 1300$	$t_m(150) = 0.0163d + 4.70$	Off, one decimal
300	$3 \leq d < 50$	$t_m(300) = 0.080d + 2.29$	Off, one decimal
300	$50 \leq d \leq 100$	$t_m(300) = 0.030d + 4.83$	Off, one decimal
300	$100 < d \leq 1300$	$t_m(300) = 0.0334d + 4.32$	Off, one decimal
600	$3 \leq d < 25$	$t_m(600) = 0.090d + 2.54$	Off, one decimal
600	$25 \leq d \leq 50$	$t_m(600) = 0.060d + 3.30$	Off, one decimal
600	$50 < d \leq 1300$	$t_m(600) = 0.06777d + 2.54$	Off, one decimal
900	$3 \leq d < 25$	$t_m(900) = 0.160d + 2.29$	Off, one decimal
900	$25 \leq d \leq 50$	$t_m(900) = 0.060d + 4.83$	Off, one decimal
900	$50 < d \leq 1300$	$t_m(900) = 0.10449d + 2.54$	Off, one decimal
1500	$3 \leq d \leq 1300$	$t_m(1500) = 0.18443d + 2.54$	Off, one decimal
2500	$3 \leq d \leq 1300$	$t_m(2500) = 0.34091d + 2.54$	Off, one decimal
4500	$3 \leq d \leq 1300$	$t_m(4500) = 0.78488d + 2.54$	Off, one decimal
Valve pressure class, P_c	**Inside diameter (d), inch**	**Inch equation t_m (inch)**	**Round**
150	$0.12 \leq d < 2$	$t_m(150) = 0.064d + 0.092$	Off, two decimals
150	$2 \leq d \leq 4$	$t_m(150) = 0.020d + 0.18$	Off, two decimals
150	$4 < d < 50$	$t_m(150) = 0.0163d + 0.185$	Off, two decimals
300	$0.12 \leq d < 2$	$t_m(300) = 0.080d + 0.09$	Off, two decimals
300	$2 \leq d \leq 4$	$t_m(300) = 0.030d + 0.19$	Off, two decimals
300	$4 < d < 50$	$t_m(300) = 0.0334d + 0.17$	Off, two decimals
600	$0.12 \leq d < 1$	$t_m(600) = 0.090d + 0.10$	Off, two decimals
600	$1 \leq d \leq 2$	$t_m(600) = 0.060d + 0.13$	Off, two decimals
600	$2 < d < 50$	$t_m(600) = 0.06777d + 0.10$	Off, two decimals
900	$0.12 \leq d < 1$	$t_m(900) = 0.160d + 0.09$	Off, two decimals
900	$1 \leq d \leq 2$	$t_m(900) = 0.060d + 0.19$	Off, two decimals
900	$2 < d < 50$	$t_m(900) = 0.10449d + 0.10$	Off, two decimals
1500	$0.12 \leq d \leq 50$	$t_m(1500) = 0.18443d + 0.10$	Off, two decimals
2500	$0.12 \leq d \leq 50$	$t_m(2500) = 0.34091d + 0.10$	Off, two decimals
4500	$0.12 \leq d \leq 50$	$t_m(4500) = 0.78488d + 0.10$	Off, two decimals

Equation 4.1 Minimum valve wall thickness calculation based on ASME B16.34 (an alternative method to ASME B16.34 Tables 4.1 and 4.3 mandatory Appendix VI)

$$t = 1.5 * \left[\frac{P_C d}{2S_F - 1.2P_C} \right]$$

where:

t: Calculated thickness (mm/inch);
P_C: Pressure class designation number (e.g., for class 1500, $P_C = 1500$);
d: Inside diameter of the valve (mm/inch);
S_F: Stress-based constant equal to 7000.

The wall thickness of the 30″ CL1500 pipeline valve with a minimum internal diameter of 625.3 mm is calculated as follows as per Equation 4.1:

$$t = 1.5 * \left[\frac{1500 * 625.3}{2 * 7000 - 1.2 * 1500} \right] = \frac{1406925}{12200} = 115.32\,\text{mm}$$

This result indicates that the wall thickness of the pipeline valve in this example is 2.54 mm less according to the Equation 4.1 calculation method than the wall thickness value calculated by the mandatory Appendix VI. Note that using the ASME B16.34 standard significantly increases the wall thickness and increases the valve's weight. The main purpose of this section is to suggest using ASME Sec. VIII Div.02 instead of ASME B16.34 for pipeline valve wall thickness calculation to reduce the pipeline valves' weight. The next step is to calculate pipeline valve wall thickness based on ASME Sec. VIII Div.02 according to Equation 4.2.

Equation 4.2 Valve wall thickness calculation as per ASME Sec. VIII Div.02

$$t = \frac{\text{D}}{2}\left(e^{\frac{p}{s}} - 1 \right)$$

t = valve thickness (inch);
D = valve diameter (inch) =30″;
p = valve design pressure (psi) = 250 barg = 250 × 14.,5 = 3625 psi;
S = allowable stress (psi).

Note: For ASTM A216 WCB carbon steel, body material of the valve = 20,000 psi at ambient temperature of 25°C as per Table A.1 of ASME B31.3.

$$e = 2,7182$$

FIGURE 4.2 30" Class 1500 pipeline valve thickness and weight comparison based on ASME Sec. VIII and ASME B16.34. (Photo by the author.)

Having the values in Equation 4.2 → t = 2.98 inch =75.71 mm (valve thickness)

Figure 4.2 illustrates and compares the weight and thickness of a 30" Class 1500 ball valve that is designed based on both ASME B16.34 and ASME Sec. VIII Div.02.

As illustrated in Figure 4.2, the valve's weight can be reduced from 23 tons to 14 tons by switching to ASME Sec. VIII Div.02 calculation for wall thickness. Saving the wall thickness on pipeline valves results in a reduction in the height and length of the valve. The next section explains how to validate the wall thickness of the valve that is calculated based on ASME Sec. VIII Div.02 through stress analysis.

4.2.1.2 Wall Thickness Validation

4.2.1.2.1 Finite Element Analysis

If the wall thickness is calculated in accordance with ASME Sec. VIII Div.02, then it is typically validated by finite element analysis (FEA) and/or by actual testing. FEA is a computerized method of predicting how components will respond to forces, stresses, and other physical effects such as flow or heat. The valve wall thickness values given by ASME B16.34 are not conservative and are accepted by valve engineers without any validation. However, the valve supplier must demonstrate that the valve functions properly under the most severe conditions and design loads provided by the customer by FEA or testing. FEA design validation is performed by valve engineers on a variety of valve components, including the body, bonnet, ball, seats, bolts, pup, and transition pieces, but only the body and bonnet require FEA for wall thickness verification. The following loads, pressures, and temperatures should be considered:

- Design pressure (DP)
- Hydrotest pressure (HP) = 1.5*DP
- Operating pressure (OP)
- Maximum design temperature (DP)
- Maximum operating temperature (OP)
- End effect due to pressure (EE)
- Axial forces (AF)
- Transverse bending moment (BM_z)
- Vertical bending moment (BM_y)
- Torsion (TOR)
- Wave loads (WL)
- Blast loads (BL)

Figures 4.3 and 4.4 illustrate axial and torsion forces applied to the valve ends from the pipeline, respectively. As illustrated in Figure 4.5, the transverse and vertical bending moments, BM_z and BM_y, are also applied to the end of a pipeline valve. In the following section, we will discuss how loads can be combined for valve stress calculations during design validation.

4.2.1.2.1.1 Load Combination and Scenarios In five scenarios, pipe loads are taken into consideration, including design, operating, hydrostatic, maintenance, and accidental. The design and operating conditions include both maximum and minimum pressure and temperature values and their range. Maximum and minimum design temperature values represent the maximum and minimum temperatures to which an item of equipment may be exposed. For example, a maximum design temperature could be ten degrees centigrade higher than a maximum operating temperature. An operating temperature can be defined as the temperature inside a facility (e.g., an industrial valve) during its working period. Pressure testing is a method used to test equipment, facilities, and components such as piping and valves. The purpose

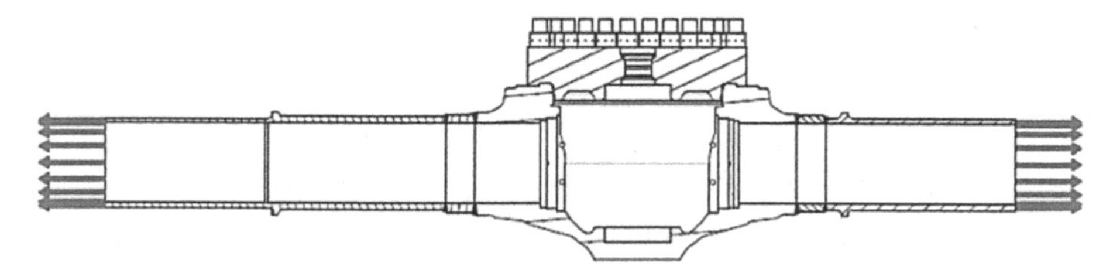

FIGURE 4.3 Applied axial forces to the pipeline valve ends.

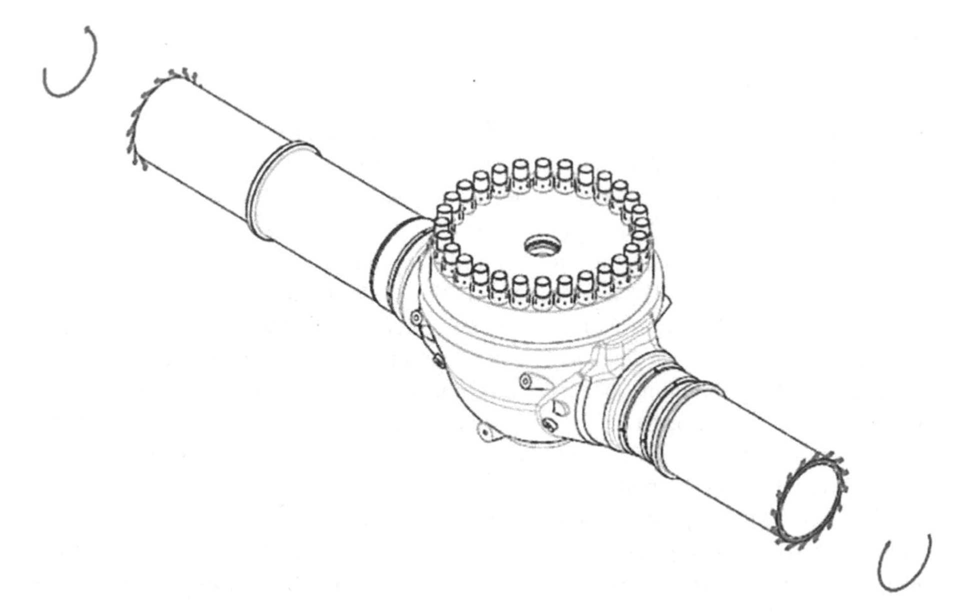

FIGURE 4.4 Applied torsion forces to the pipeline valve ends.

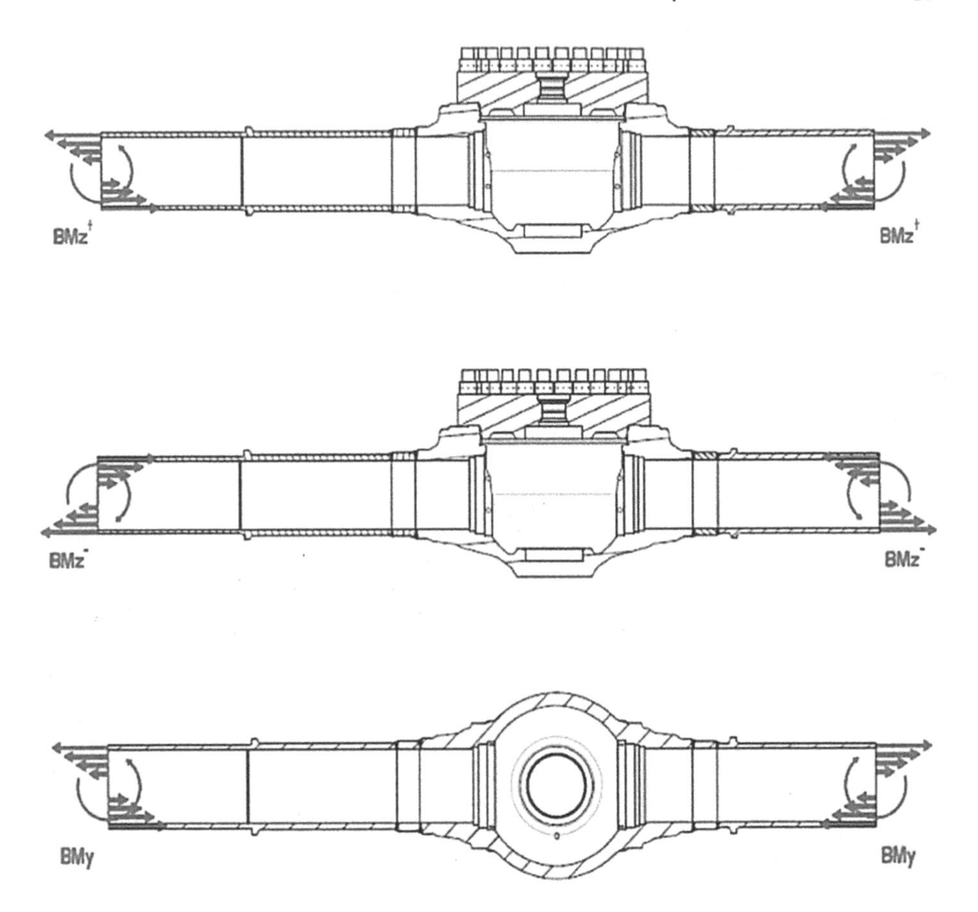

FIGURE 4.5 Applied transverse (BM_z) and vertical (BM_y) bending moments to the pipeline valve ends. (Photo by the author.)

of this method is to assess the structural integrity of equipment and components as well as detect possible leakages from them. A hydrostatic test is typically conducted by using fresh, clean water, which is free of solids and particles. There should be a maximum of 20, 30, or 50 parts per million of chloride in this water. Especially for austenitic stainless steel and carbon steel, reducing the chloride content of the water makes more sense. The reason behind this is that they are more prone to pitting and chloride stress cracking corrosion. If the test water is expected to be corrosive, it is generally recommended to add less than 1% corrosion inhibitor to the test water. Ideally, the piping should be free from dirt, particles, scale, and rust and allow water or air to flow freely through it. A general rule of thumb, test pressure is 1.5 times the internal design pressure (DP). In a maintenance case, the bonnet is removed from the valve's body, leaving the body exposed to more stress. Oil and gas production lines are shut down during pipeline maintenance. The term "accidental" or "occasional" refers to loads applied to the piping system during a given period or part of a total operating period. In the event of an explosion offshore, the blast load or wave acts as the main accidental load. The blast load can be assumed to be equal to 0.3 bar as a drag force, or the value in the project specifications given in terms of gravity acceleration parameter "g." Offshore structures such as platforms and topside facilities are subjected to wave loads (WL) influenced by parameters such as wave height and period.

In general, the highest loads on pipeline valves occur during accidental and hydrostatic conditions, while the lowest loads occur during maintenance and operation. At design conditions, pipeline valve loads, such as axial, torsion, and bending, are lower than those for accidental and hydrostatic scenarios, and higher than those during operations and maintenance. Equations 4.3, 4.4, 4.5, 4.6, and 4.7 are used to combine the pressure and load values applied to offshore pipeline valves during design, hydrostatic testing, operation, maintenance, and accidental conditions, respectively.

Equation 4.3 Design load combination for offshore pipelines valve

$$\text{Design load combination} = \text{DP} + \text{EE} + \text{AF} + \text{TOR} + BM_y + BM_Z + \text{WL}$$

DP is the main load that is considered in the design scenario. Together, the combined design loads result in axial, torsion, and vertical loads, as well as transverse and vertical bending moments. Considering the author's industrial experience, WL is only considered in the design load combination scenario in this example. Furthermore, end effect load due to pressure (EE) is considered as a part of the design load.

Equation 4.4 Hydrostatic test load combination for offshore pipeline valves

$$\begin{aligned}\text{Hydrostatic load combination} &= \text{HP} + \text{EE} + \text{AF} + \text{TOR} + BM_y + BM_Z \\ &= 1.5 * \text{DP} + \text{EE} + \text{AF} + \text{TOR} + BM_y + BM_Z\end{aligned}$$

Equation 4.5 Operating load combination for offshore pipeline valves

$$\text{Operating load combination} = \text{OP} + \text{EE} + \text{AF} + \text{TOR} + BM_y + BM_Z$$

Equation 4.6 Maintenance load combination for offshore pipeline valves

$$\text{Maintenance load combination} = \text{AF} + \text{TOR} + BM_y + BM_Z$$

Equation 4.7 Accidental load combination for offshore pipeline valves

$$\text{Accidental load combination} = \text{BL} + \text{OP} + \text{EE} + \text{AF} + \text{TOR} + BM_y + BM_Z$$

Modeling and analyzing the valve in FEA are next steps. These will be discussed in the following section.

4.2.1.2.1.2 Modeling and Analysis An engineer can use ANSYS software (e.g., workbench version 17.1) for modeling and analysis of offshore pipeline valves. Regarding the modeling details, the materials can be modeled as described in ASME BPVC VIII. Div.02. The valve analyzer can select tetrahedral or hexahedral elements (e.g., 20-node second-order SOLID 186) to achieve better numerical results. SOLID 186 is a higher-order 3D 20-node solid element from the ANSYS reference manual that exhibits quadratic displacement behavior. The element is defined by 20 nodes having three degrees of freedom per node: translations in the nodal x, y, and z directions. Figure 4.6 shows the meshes and criticality of the loads applied to a pipeline valve. The principal analysis includes Von Mises stresses and deformation measurement in X, Y, and Z directions.

The stresses calculated through Equations 4.4, 4.5, 4.6, and 4.7 are linearized or converted through the thickness with Von Misses stress theory into two types: membrane stress and bending stress. Membrane stress is the average stress across the thickness of a component. Bending stress is variable across the thickness of the component and is made of compression and tensile stresses applied to the component's longitudinal axis. Figure 4.7 illustrates the distribution of membrane stress (P_m *or* P_L),

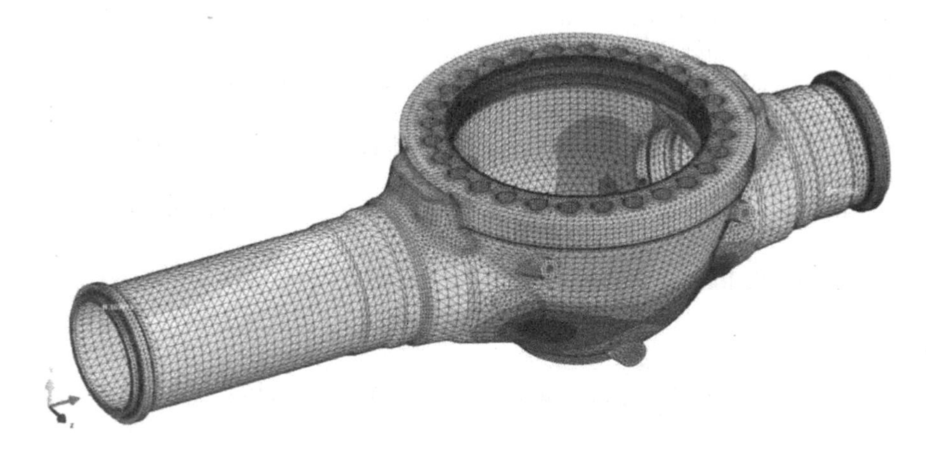

FIGURE 4.6 Meshes and applied loads criticality on a pipeline valve.

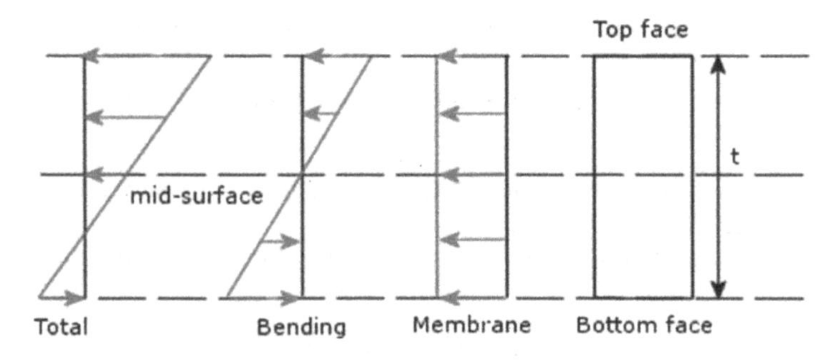

FIGURE 4.7 Distribution of membrane, bending, and bending plus membrane stress across the thickness of a component. (Photo by the author.)

bending stress (P_B), and bending plus membrane stress (P_{m+b}) across the thickness of a component. Peak stress is defined as the highest stress, which can be ignored in FEA in this case. To be more specific, there are two types of membrane stresses: a general membrane and a local membrane. General membrane stresses are found away from critical locations like supports, sudden changes in size or cross-section, geometry transformation, and irregularities. They are compared directly with the allowable stresses, parameter S_a. Local membrane stresses (P_L) apply higher stress levels at critical areas, so they are limited to 1.5 times the allowable stresses, or $1.5*S_a$. Same as the limitation for local membrane stresses, bending plus membrane stress shall not exceed 1.5 times the allowable stresses. The limits of FEA linearized stresses are summarized in Equation 4.8.

Equation 4.8 Limits of FEA linearized stresses

$$P_m < S_a$$

$$P_L < 1.5 * S_a$$

$$P_m + P_b < 1.5 * S_a$$

$$P_L + P_b < 1.5 * S_a$$

The most pertinent consideration is to model the valve body and bonnet materials and mechanical characteristics at a minimum and maximum design condition for the design load combination. In addition, the materials' mechanical properties for operating and accidental conditions are obtained by using both minimum and maximum operating temperatures. Finally, the materials are subjected to ambient temperatures during hydrostatic and maintenance tests. Increasing the temperature generally reduces mechanical properties, such as yield strength, tensile strength, and allowable stress.

Example 4.1

The body of a pipeline ball valve is made of carbon steel, ASTM A216WCB. The valve's minimum and maximum design temperature values are -10°C and 200°C, respectively. The tensile, yield, and allowable stress values of the carbon steel valve body at 200°C are 480, 209, and 152 MPa. The applied stresses on the valve are linearized through the valve thickness in Section 1, as shown in Figure 4.8. The maximum values of the general membrane, local membrane, and bending membrane are 70, 120, and 120 Mpa, respectively. Will the thickness of the valve be adequate to withstand the applied loads at 200°C?

Answer) All conditions provided in Equation 4.8 shall be satisfied at 120°C.

$$P_m < S_a \rightarrow 70 < 152 \text{OK}$$

$$P_L < 1.5 * S_a \rightarrow 120 < 1.5 * 152 = 228 \text{OK}$$

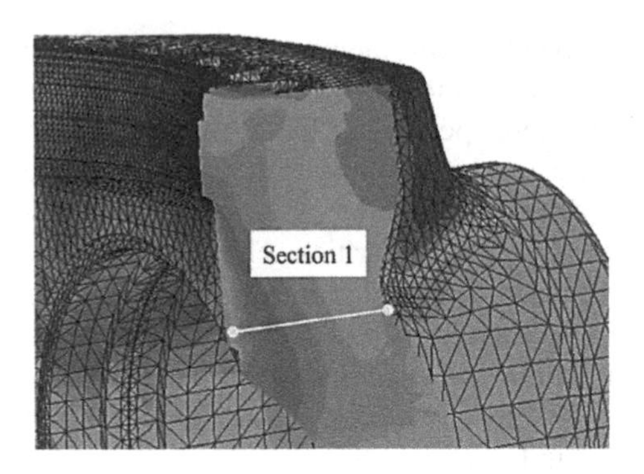

FIGURE 4.8 Stress intensities through the thickness of a pipeline ball valve highlighted as Section 1. (Photo by the author.)

$$P_m + P_b < 1.5 * S_a \rightarrow 70 + 120 = 190 < 1.5 * 152 = 228 \text{ OK}$$

$$P_L + P_b < 1.5 * S_a \rightarrow 120 + 120 = 240 > 1.5 * 152 = 228 \text{ (not satisfied)}$$

As a result of not meeting the last condition, the thickness of the valve in Section 1 is not adequate to withstand applied loads; therefore, the valve engineer should increase the thickness of the valve.

4.2.1.2.2 Load Test (Bending Moment Test)

The second strategy to verify the suitability of a pipeline valve wall thickness against applied loads is to conduct a load test called a bending moment test, which can be found in texts such as NORSOK L-002, piping design, layout, and stress analysis standard. The valve manufacturer is responsible for implementing this test. Norwegian Oil and Gas standards NORSOK are developed to ensure adequate safety, value-adding, and cost-effectiveness for developments and operations in the petroleum industry. As far as possible, NORSOK standards are intended to replace the specifications of the oil companies and serve as references for the authorities' regulations. Figure 4.9 illustrates a load or bending moment test for an offshore pipeline ball valve. The previously discussed loads are applied to the valve, and electrical strain gauge sensors are trapped in several critical valve locations. It is the primary function of the gauges to measure the exerted force and potential displacement. FEA simulations are validated or rejected based on monitored and measured parameters by gauges. As a result of the applied loads during the test, the valve body may be damaged internally or externally. A deformed body can cause the seat to leak from within. An internally deformed body may damage the seat and ball, preventing the ball from being removed from the valve during maintenance. It was explained in Chapter 1 that the top-entry design used for offshore pipeline valves provides more load resistance than the split or side-entry designs. Tests can be repeated for a variety of load cases, including design, operation, hydrostatic, accidental, and maintenance.

FIGURE 4.9 Load test on an offshore pipeline valve and highlighted strain gauges on the valve body. (Photo by the author.)

It is possible that during the maintenance load test, the personnel involved with testing the valve remove the bonnet from the top of the valve, resulting in the ovality of the valve body so that an operator cannot reassemble the bonnet on the valve after the test. By adding two pup pieces at both ends to the valve length, it is possible to simulate the actual bending moment. As the bending moment depends on force and distance, increasing the distance through the addition of pup pieces can lead to higher bending moments required for the test.

An end user may request the bending test in order to verify and validate the valve wall thickness has been approved by FEA analysis. On the other hand, the test can be omitted in many cases due to two main disadvantages. The first is that it may damage the valve, and the second is that this test is performed after the valve has been manufactured. In the event that the valve fails during this test, it would be very late to manufacture a new valve from scratch with adequate body and bonnet wall thickness values.

4.2.2 Cylindrical Nuts

A heavy hexagonal nut (Figure 4.10) is larger and thicker than a standard hexagonal nut and is covered by ASME B18.2.2. This type of nut is commonly used for piping and valves in the oil and gas industry. In some instances, valve manufacturers use heavy hexagonal nuts for their pipeline valves, resulting in a larger and heavier valve than when cylindrical nuts are used.

A cross-section of a cylindrical nut encircled in a hexagonal section is shown in Figure 4.11. The hexagonal section length, parameter s(hexa), is almost equal to 1.15 times the cylinder section length, parameter r(cyl), and the hexagonal section area is

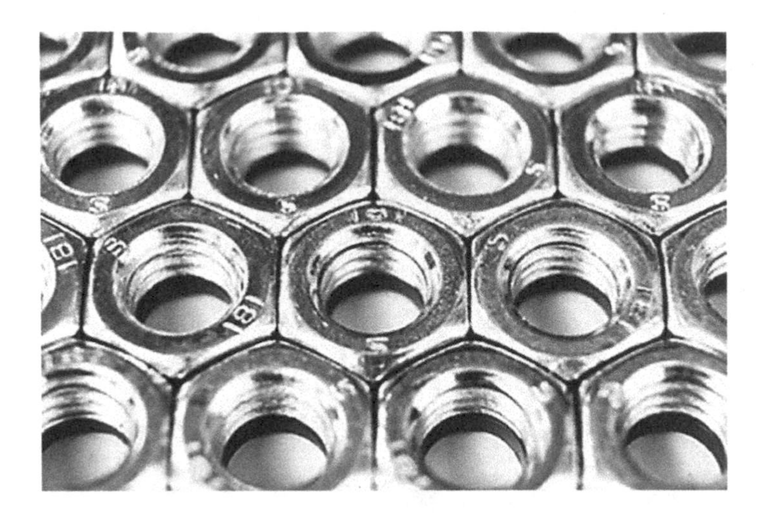

FIGURE 4.10 Heavy hexagonal nuts. (Courtesy: Shutterstock.)

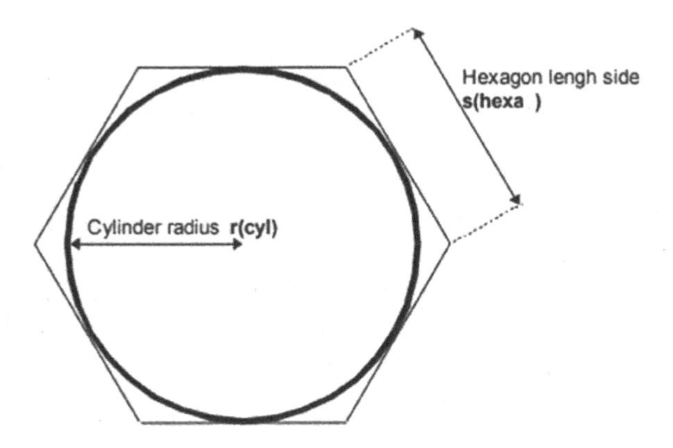

FIGURE 4.11 The cross-section of a cylindrical nut enclosing a hexagonal section.

almost identical to 1.10 times the cylinder section area. As a result, the area of one cylindrical nut is nearly 90% of that of a hexagonal nut, and there are several nuts for the body and bonnet joints. The nuts are located on the bonnet, so using cylindrical nuts makes the bonnet more compact and reduces the bonnet weight. Figure 4.12 illustrates an electrically actuated pipeline ball valve with cylindrical nuts for the body and bonnet joints in order to reduce the valve's weight.

FIGURE 4.12 A pipeline ball valve with cylindrical nuts for connecting the body and the bonnet that operates electrically. (Photo by the author.)

QUESTIONS & ANSWERS

1. What is the correct statement regarding the calculation of minimum valve wall thickness in accordance with ASME B16.34?

   A. According to mandatory Appendix VII, the wall thickness of a ball valve with an inside diameter of 840 mm and a pressure class of 1500 is 157.5 mm.
   B. Table 4.1 taken from ASME B16.34, which is used for the calculation of wall thickness, has the drawback that it does not cover all possible internal diameters of valves. Therefore, it is not possible to provide wall thickness values for internal diameters that are not covered.
   C. ASME B16.34 is the correct reference for the wall thickness calculation of the pipeline valves.
   D. Valves that have the same internal diameter and pressure class but are made of different materials have different body wall thicknesses according to the ASME B16.34 standard.

   Answer) Option A is correct. Let's try calculating wall thickness according to the basic equations provided in mandatory Appendix VI in ASME B16.34:

   P_C = Pressure class 1500 and d = 840mm, $3 < d \leq 1300 \rightarrow t_m(1500) = 0.18443d + 2.54 = 0.18443*840 + 2.54 = 157.46$ mm~ 157.5 mm.

   Option B is not completely correct: Although not all of the internal diameters are not covered by ASME B16.34, it is possible to interpolate the valve wall thickness values associated with the missing internal diameters. Option C is incorrect because the pipeline valve wall thickness is calculated according to ASME Sec. VIII Div.02 to save weight. In accordance with the ASME B16.34 standard, option D is not correct because the type of material and its mechanical strength do not affect the valve's wall thickness according to ASME B16.34. As long as they have the same internal diameter and pressure class, valves made of different materials will have the same wall thickness.

2. Figure 4.13 illustrates a pipeline ball valve with two wall thickness calculation methods, one on the top and the other at the bottom, designed according to ASME B16.34 and ASME Sec. VIII Div.02. Which statement about the valves is correct?

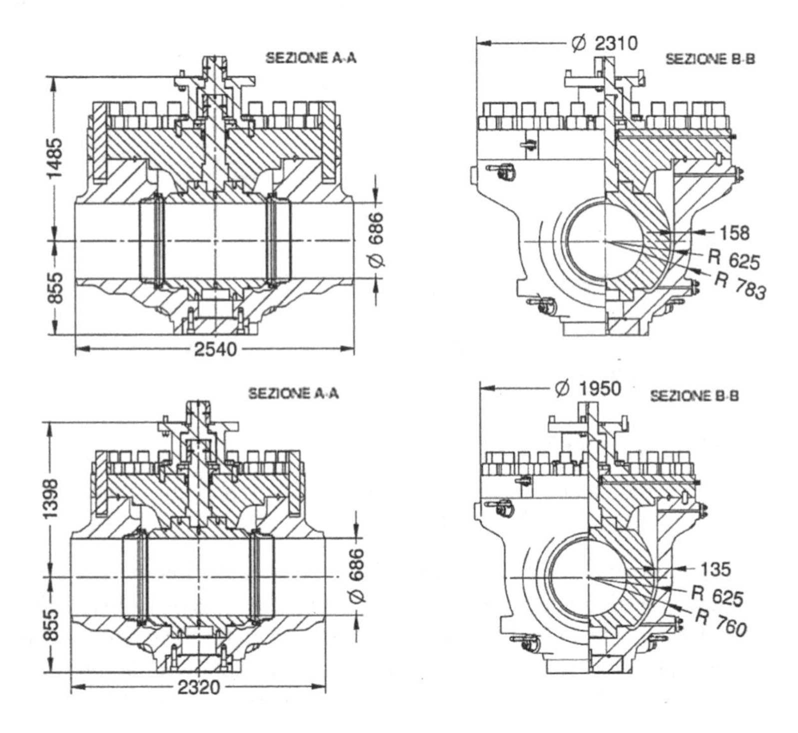

FIGURE 4.13 A pipeline ball valve designed based on two different codes and standards.

A. For the top valve, ASME Sec. VIII Div. 02 is used to calculate the wall thickness.
B. There is no effect of the wall thickness on the valve's height or end-to-end measurements.
C. The internal diameter of the valve is affected by the wall thickness of the valve.
D. The valve at the bottom is lighter than the one on the top.

Answer) Option A is incorrect because the valve at the top is thicker than the valve at the bottom. Thus, wall thickness of the valve at the top is calculated according to ASME B16.34. Option B is incorrect since the thicker valve is longer end-to-end and has a higher height. Option C is not valid, as the wall thickness of the valve does not affect its internal diameter. The answer is D since the valve at the bottom has a thinner wall thickness, making it lighter.

3. What is the correct statement regarding the valve wall thickness validation calculated according to ASME Sec. VIII Div.02?

   A. An FEA analysis is conducted only on both the balls and seats.
   B. The valve engineer takes into account only the load scenarios for the design and operation of the valve.
   C. The valve analysis includes various forces applied from the connected pipeline, such as axial, torsion, bending, as well as wave and blast forces.
   D. The peak stress is checked against the allowable stress after stress linearization.

   Answer) As the FEA analysis for wall thickness validation is applied to the body and bonnet, option A is incorrect. Likewise, option B isn't correct because in addition to design and operating scenarios, loads during the hydrostatic test, accidental or occasional loads like blasts, and loads during maintenance are taken into account. The correct choice is C. FEA analysis ignores peak stresses, so option D is incorrect.

4. What is the main purpose of bending moment or bending load test?

   A. Validating the valve design against internal pressure and leakage
   B. Validating the valve body and bonnet wall thickness against the loads
   C. Proving the integrity between the valve and actuator
   D. Proving the wear resistance of the valve after several cycles (opening and closing)

   Answer) Option B is the correct answer.

5. Find the wrong sentence regarding the body and bonnet joints' bolts and nuts for the pipeline valves.

   A. Heavy hexagonal nuts designed according to ASME B18.2.2 make the valve heavier than cylindrical nuts.
   B. Using cylindrical nuts reduces the weight of the valve's bonnet.
   C. Using cylindrical nuts is the only way to reduce the weight of pipeline valves.
   D. Cylindrical nuts studied in this chapter have less area than hexagonal nuts.

   Answer) Option C is incorrect, as the calculation of pipeline valve wall thickness based on ASME Sec. VIII instead of ASME B16.34 is another way of reducing the weight of pipeline valves.

BIBLIOGRAPHY

1. Alchalabi M., Leboeuf R. and Sherertz C. (2010). Light-weight topsides for heavy-weight projects. Offshore. [online] available at: https://www.offshore-mag.com/business-briefs/equipment-engineering/article/16763863/lightweight-topsides-for-heavyweight-projects [access date: 18th of December 18, 2021].
2. American Society of Mechanical Engineers (ASME). (2012). Design and Fabrication of Pressure Vessels. Boiler and Pressure Vessel Code. ASME Section VIII Div.02. New York, NY. USA.

3. American Society of Mechanical Engineers (ASME) B 18.2.2 (2015). Nuts for general applications: Machine screw nuts, hex, square, hex flange and coupling nuts (inch series) New York, NY. USA.
4. American Society of Mechanical Engineers (ASME) B16.34. (2017). *Valves–Flanged, threaded, and welding end.* New York, NY: ASME.
5. Audubon (2021). Importance of offshore weight control. [online] available at: https://auduboncompanies.com/importance-of-offshore-weight-control/ [access date: 18th of December 18, 2021].
6. Boyd N.G. (1986). Topsides weight reduction design techniques for offshore platforms. Proceeding paper presented at the offshore technology conference, Houston, Texas. Paper Number: OTC-5257-MS.
7. Carter G. (2001). Offshore platform rigs adapting to weight-space restrictions for floaters. Offshore. [online] available at: https://www.offshore-mag.com/drilling-completion/article/16763331/offshore-platform-rigs-adapting-to-weightspace-restrictions-for-floaters [access date: 18th of December 18, 2021].
8. Det Norske Veritas (DNV) GL-CG-0130 (2018). *Wave loads.* Hovik, Norway: DNV.
9. Grasp Engineering (2020). What is stress linearization? [online] available at: https://www.graspengineering.com/what-is-stress-linearization/ [access date: 22th of December 18, 2021].
10. NORSOK L-002 (1997). *Piping design, layout and stress analysis*, (2nd revision). Lysaker, Norway.
11. Prigniau M. et al. (2021). Difference between hexagonal prismatic and cylindrical cell for infinite array calculation. [online] available at: https://www.irsn.fr/EN/Research/publications-documentation/Publications/DSU/SEC/Documents/NCSD05miss_.pdf [access date: 22th of December 18, 2021].
12. Sotoodeh K. (2015). Top entry export line valves design considerations. *Valve World Magazine*, 20(05), 55–61.
13. Sotoodeh K. (2018). Pipeline valves technology, material selection, welding, and stress analysis (A case study of a 30 in class 1500 pipeline ball valve). *American Society of Mechanical Engineers (ASME), Journal of Pressure Vessel Technology*, 140(4), 044001. https://doi.org/10.1115/1.4040139. Paper No. PVT-18-1043.
14. Sotoodeh K. (2022). *Coating application for piping, valves and actuators in offshore oil and gas industry.* (1st edition). Oxford, UK: CRC Press.
15. Sotoodeh K. (2022). *Cryogenic valves for liquified natural gas plants*, (1st edition). Cambridge, The USA: Elsevier (Gulf Professional Publishing).
16. Sotoodeh K. (2022). *Piping engineering: Preventing fugitive emission in the oil and gas industry*, (1st edition). New York, USA: Wiley.

5 Seat Scrapers and Flushing Ports

5.1 INTRODUCTION

The entrapment of oil, debris, and particles inside the seats of pipeline valves is one of the most common causes of failure. Figure 5.1 illustrates the arrangement of the ball valve seat, including the components. Item #6 is the seat that is in contact with the ball on the right side. For effective and tight sealing, item #700 is a spring that pushes the seat toward the ball. Item #522 is a seal between the valve's body and its seat.

Crude oil and gas in the export pipeline are typically treated and cleaned. Under certain conditions, the crude oil may contain dissolved waxes that precipitate and deposit. Oil, for example, may accumulate in the seat arrangement over time and become harder and waxy. It is possible that hardened oil or wax might impair the seat's functionality and tightness toward the ball, which may result in internal leaks. When the seat is blocked in its position and cannot retract due to the wax effect behind the seat, the condition can worsen. In such a scenario, the seat could provide excessive loading on the ball-contact surfaces, causing galling and excessive wear to these components. A second undesirable consequence of the wax deposition is the inability to remove the ball from the valve for maintenance, as the seat exerts a great deal of force. If this problem persists, ***flushing ports*** can be used to inject some solvent or steam into the seat arrangement to dissolve the wax, including the areas at the back of the seat.

Newly constructed pipelines contain debris in the form of welding rods, weld slags, mill scale, and corrosion. The buildup of debris can threaten the integrity and safety of the pipeline, which is why pigs (pipeline inspection gadgets) are used to remove the debris before the pipeline is started and operated. Solid particles, dirt, or debris can enter the seat arrangement from the back and clog or damage the seat. For example, seals between the seat and body, such as item #522, are at risk of damage by particles. Thus, a ***seat scraper*** (item #530) made of Teflon or Viton® materials

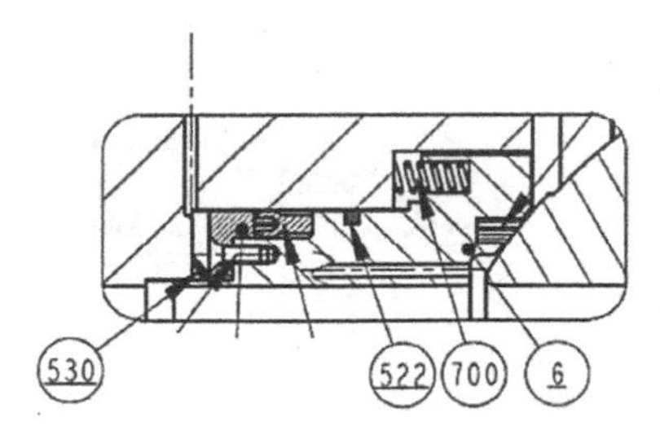

FIGURE 5.1 A ball valve seat arrangement supplied with a seat scraper. (Photo by the author.)

DOI: 10.1201/9781003343318-5

prevents the ingress of particles to the seat. The valve manufacturer and end user can consider a seat scraper for both metal and soft- (non-metal) seated valves.

5.2 WAX DISSOLUTION IN SEATS

5.2.1 Sealant Injection Fittings

Sealant injection ports, points, or fittings are designed to inject sealants both on the seats and on the stem. Valve operation can result in damage to the stem and seat seals. For this reason, sealant is injected through the sealant injection fitting for the emergency repair of seals or any other seat damage. The emergency sealant injections are installed on both soft and metal seats to repair the seats and delay maintenance on damaged seats until a pre-scheduled maintenance time is available. Figure 5.2 illustrates that the injection point is comprised of a check valve (item #43) screwed to the valve's body, an injection fitting (item #41), and finally, item #552, a gasket or seal. Figure 5.3 illustrates a pipeline valve with eight sealant injection ports on seats, four points on each seat, in the size of ¾". In some cases, it is common to apply a small injection fitting in order to inject a solvent or steam. By using this method, the wax volume accumulated at the back of the seat or other places in the seat arrangement, as well as the contact areas between the seat and ball, will be removed. The injection fittings are, however, small and have inline check valves which may reduce flow efficiency and prevent complete and effective removal of wax deposits. In this case, the alternative solution is to use ***flushing ports***, as described in the following section.

5.2.2 Flushing Ports

In the author's experience, an end user expressed concern that having separate flushing ports on offshore pipeline valves might cause increased leakage risk due to more penetrations on the valve bodies. A valve manufacturer proposed flushing ports that are larger than sealant fittings to improve the injection efficiency and cleaning. As shown in Figure 5.3, two flushing ports are installed on each seat, totaling four flushing ports. As shown in Figure 5.4, the sealant injection and flushing ports on the seats of a pipeline ball valve can be seen more clearly. Each seat is equipped with two 1" flushing ports that are horizontally oriented and larger in diameter than the sealant injection ports in order to allow a high flow rate for efficient cleaning. Flushing ports are integrally incorporated with a 1" flange in pressure class 1500 on the outer part,

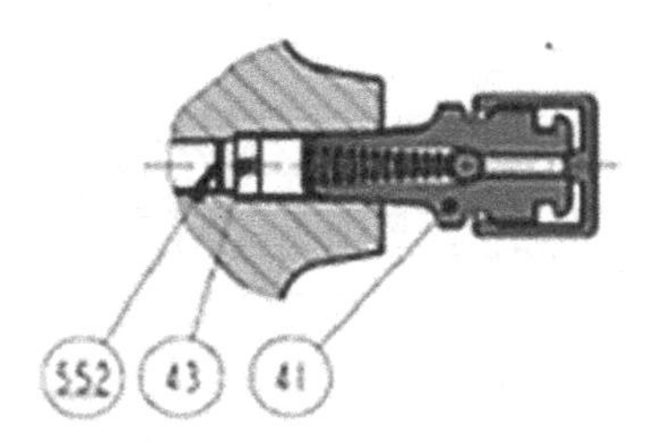

FIGURE 5.2 A sealant injection port. (Photo by the author.)

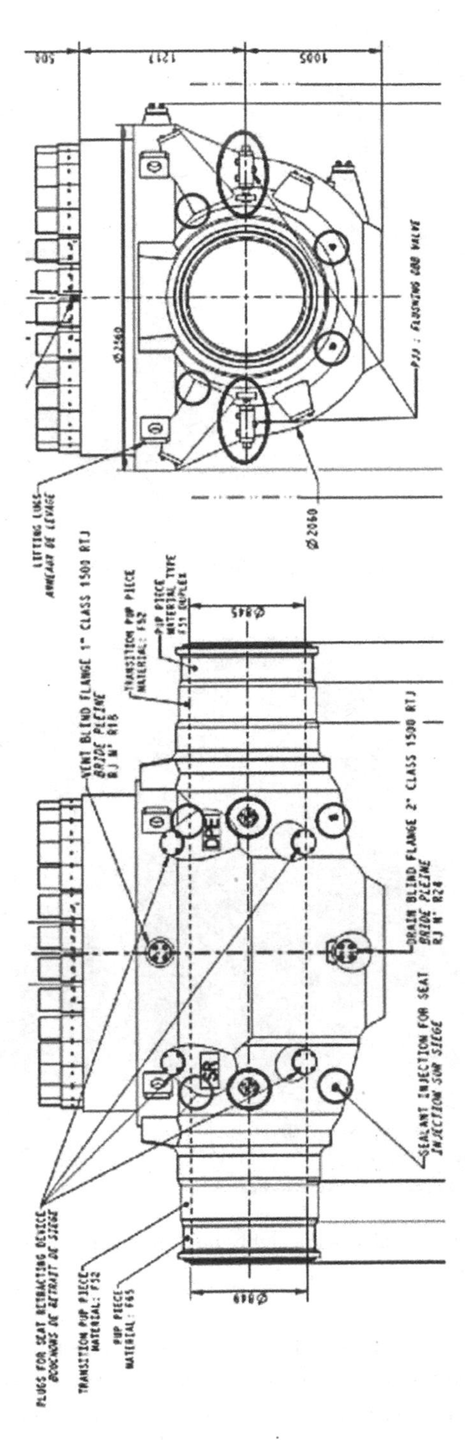

FIGURE 5.3 A pipeline valve with eight sealant injection ports on seats and four flushing ports.

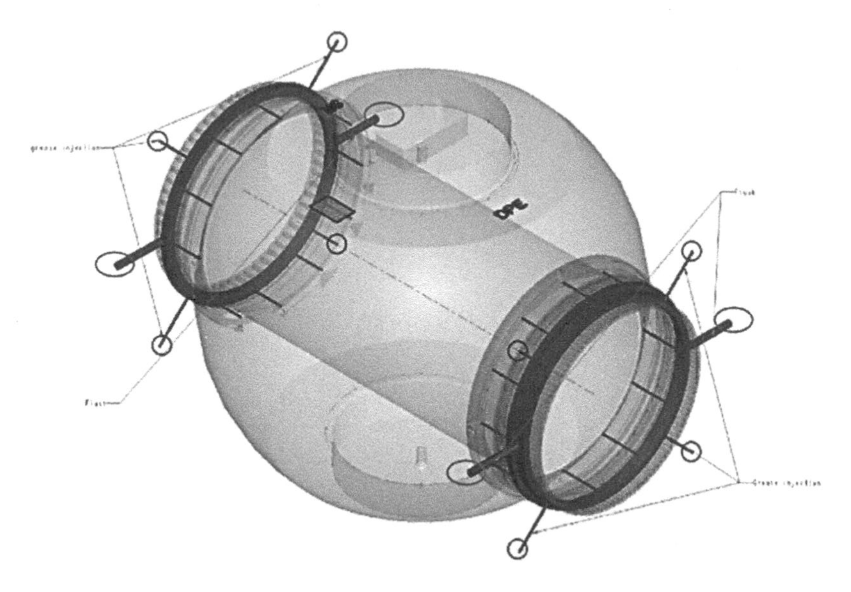

FIGURE 5.4 A view of sealant injection and flushing ports on the seats of a pipeline valve.

FIGURE 5.5 A pipeline ball valve with highlighted modular valves installed on flushing ports. (Photo by the author.)

which is also plugged by a modular valve, consisting of two ball valves and a needle valve between them (see Figure 5.5) to provide double isolation of the internal fluid from the external environment. Because of the high-pressure conditions in which the valve operates, double isolation must be provided by a modular valve rather than a single valve. As modular valves are normally closed, an operator should open them during the flushing operation. As with sealants, flushing fluid is injected through the holes and channels drilled in the seats. Waxes deposited and entrapped at the backside of seats dissolve in the flushing fluid and are carried to the piping system. Surfaces and volumes of the backseat area should be large enough to allow filling and accumulation of flushing fluid pressure to flush out the hardened oil. The flow and pressure rates of the flushing fluid should be increased in order to achieve higher cleaning efficiency; these rates are dependent on various parameters such as gravity forces, the viscosity of the flushing fluid, and the valve's internal pressure.

5.2.3 Flushing Fluid

Under normal operating conditions, any fluid capable of dissolving wax is acceptable, provided two conditions are met. First, the flushing fluid should be diluted with the production oil. Second, the end user receives and approves the material safety datasheet. Material safety datasheets provide information related to occupational safety and health for products and substances. Steam and kerosene are two common flushing fluids used at temperatures below 200°C. Fluid flushing temperatures should be limited in order to avoid damaging soft seals, such as lip seals and Teflon scrapers, in seat arrangements.

5.2.4 Flushing Procedure

The flushing operation is performed in two different ways: the first, known as a ***normal flushing operation***, involves injecting flushing fluid at the operating pressure into the depressurized and empty valve cavities. Second, there is a ***special flushing operation*** in which the flushing fluid is injected at a pressure higher than the operating temperature when the valve cavities are pressurized and filled with oil. In both conditions, the operator uses a suitable pump connected to the outlet of the modular valve. There is a check valve installed between the pump and the modular valve in order to prevent backflow of oil or flushing fluid from the inside to the outside. One method of cleaning flushing ports is to open modular valves when the pipeline valves' cavities are under pressure. This can replace oil at the back of the seats with fresh oil. However, this method is not recommended because of the danger of high-pressure fluids being discharged into the environment. An alternative solution is to inject the flushing fluid from one port and collect it from the opposite port without any pressure in cavities, which allows the flushing port channels to be cleaned. Flushing frequency varies with the oil properties and the degree of wax deposition; however, a valve supplier recommended six months intervals for flushing as a baseline. Operators may prefer shorter flushing intervals, but it is essential to keep the seats free of hardened oil or waxy products. Cleaning is most effective when both flushing ports on one seat are connected to pumps and both pumps are injected simultaneously. In normal flushing operations, operators increase the injection pressure and flow of pumps to the operating pressure, or to a pressure higher than the operating pressure in special flushing operations. Depending on the amount of wax deposited in the valve seat, the size of the valve, and other parameters, this operation may take several hours with steam, or a longer time with kerosene. Following simultaneous injection through both ports on a seat, the operator can flush through one port while the other is closed and then switch to the opposite condition.

5.3 PARTICLES INGRESS PREVENTION IN SEATS

Pipeline valves are typically operated in clean fluid services. Nevertheless, debris and particles such as welding rods, weld slags, mill scales, and rust remain in the pipeline after installation. It is possible that the pipeline valves will be contaminated with debris and particles during pre-commissioning. During pre-commissioning, a number of activities are performed, such as cleaning and inspecting internal dimensions, which prepare the pipeline and connected valves for operation. Solid particles, dirt, and debris may enter the seat arrangement from the back and cause the seat to clog or become damaged. At the back of the seats, a ***seat scraper*** made from Teflon or Viton® prevents particles from entering. Seat scrapers are not limited to soft-seated valves. In general, seat scrapers are recommended for valves that operate in dirty or particle-containing fluid services.

QUESTIONS & ANSWERS

1. Select the correct sentence regarding seat protection against the particles.

   A. By selecting a metal seat valve, you can ensure the safe operation of the valve in liquids containing particles.
   B. In order to prevent wax accumulation in the seat arrangement, valve manufacturers use a seat scraper.
   C. In clean service, valves do not require any seat protection against particles.
   D. There is a seat scraper made from Teflon material that is placed at the back of the seat to prevent particles from entering.

   Answer) It is incorrect to select option A. The valve specifications and datasheets require a metal seat for particle-containing services; however, the particles could damage the metallic seat during operation. Option B is also in error, as the seat scraper prevents particles and debris from entering the seat arrangement and not the wax. Option C is incorrect because seat protection for the valves may be needed to protect against debris and particles generated during construction of the pipeline prior to the operation. The correct option is D.

2. During the injection of the flushing fluid to the seats, why is the maximum temperature of the flushing fluid limited?

   A. Improve the cleaning efficiency
   B. Take care not to damage seals or soft materials in seat arrangements
   C. Improved dilution of the flushing fluid in the production oil fluid
   D. Combine the sealant injection and flushing ports

   Answer) Option B is the correct answer. A high temperature can melt and damage soft seals and materials other than metal, such as scrapers in the seat arrangement.

3. Which sentence is correct about flushing the seats to prevent wax deposition?

   A. Flushing operations should be performed daily.
   B. Typically, more than five flushing ports per seat are required to achieve high cleaning efficiency.
   C. The flushing fluid should be diluted in the produced oil, and it does not have any health and safety issues.
   D. The channels circulating the flushing fluid are separate from the sealant injection fittings.

   Answer) Option A is not correct as daily flushing operation is too frequent; a six-month flushing frequency as a base case is proposed by valve manufacture. Option B is not correct either, and only a maximum of two ports per seat is sufficient. Option C is the right answer. Option D is wrong because the flushing fluid and sealant distributing channels are the same.

4. What type of valve is suitable for flushing ports isolation from the environment?

   A. A modular valve with two ball valves and a needle valve in between
   B. Wedge gate valve
   C. Butterfly valve
   D. Double isolation and bleed (DIB) ball valve with two double piston effect (DPE) seats

   Answer) Option A is the correct response since flushing ports are operated at high operating pressures and at the risk of flammable hydrocarbon leakage to the environment, thus requiring double isolation. Both wedge and butterfly valves cannot provide double isolation; therefore, both options B and C are incorrect. A DIB ball valve with double DPE seats provides double isolation; however, it is not a common valve choice for a 1" (small size) flushing port, so option D is incorrect.

5. During the flushing process, what is the primary function of the check valve installed between the pump and flushing port?

   A. Improving the cleaning efficiency
   B. Preventing backflow of injection fluid and wax into the operator
   C. Providing safety of the operator during flushing fluid injection
   D. Both options B and C are correct

   Answer) Option D is correct.

BIBLIOGRAPHY

1. Aubin Group. (2018). Tips for removing debris from a pipe. [online] available at: https://www.aubingroup.com/blog/tips-for-removing-debris-from-a-pipe/ [access date: 23th of December 18, 2021].
2. Flow Control Technology (FCT). (2015). Seat flushing procedure Statoil Johan Sverdrup 38" 1500# oil export riser ball valves. Document number: RD-MA-0069-X Rev.A.
3. Sotoodeh K. (2015). Top entry export line valves design considerations *Valve World Magazine*, 20(05), 55–61.
4. Sotoodeh K. (2021). *A practical guide to piping and valves for the oil and gas industry*, (1st edition.). Austin, USA: Elsevier (Gulf Professional Publishing).

6 Manufacturing Process

6.1 INTRODUCTION

Many manufacturing stages for valves are listed in the inspection and test plan (ITP) generated by valve manufacturers. The ITP is a key document listing all of the valve manufacturer's activities, including engineering, manufacturing, inspection, and testing. In addition, an ITP contains reference procedures for each activity, acceptance criteria, generated documents, and certificates, as well as requirements for the level of inspection. The ITP consists of various parties such as a client or third-party inspector, valve's supplier, and a sub-supplier with specific inspection requirements, including review, witness, and hold points. A review point indicates that the required party can review the documents after the activity without needing to sign or approve them. In a witness point, attendance at the site is required and a signature is required on the resulting documents and checklists; however, if the required party is not present, the activity can proceed and the papers will be signed later. An inspection at the hold point is the highest level, meaning an inspector must visit onsite and sign the checklists, and if not, the activity cannot proceed. Forging and casting valves' parts, machining, welding, non-destructive testing (NDT), internal inspections and quality control, testing, final inspection, packing, and preservation are all essential manufacturing milestones. The following section provides an overview of the manufacturing processes customized for pipeline valves.

6.2 PIPELINE VALVES MANUFACTURING PROCESSES

6.2.1 Kick-off Meeting and Engineering

The kick-off meeting is the first meeting between the client and the valve and actuator manufacturers. Ultimately, the winning manufacturer or supplier for each bid is selected for the kick-off meeting, and the purchase order (PO) is signed. During the project, all parties discussed many technical and commercial points, as well as the necessary valve documentation and due dates, inspection, schedule, and the associated risks with delivering the valves. The valve lifecycle usually begins in the early engineering phases of a project, such as the basic or feed phases, whereas the kick-off meeting takes place during the detail engineering phases. The contract review and award between the valve supplier and client takes place after the kick-off meeting. As soon as the contract is signed, the valve manufacturer begins the engineering process by preparing drawings and datasheets. The valve manufacturer prepares technical and commercial proposals and submits them to the client for approval.

DOI: 10.1201/9781003343318-6

6.2.2 Purchasing or Manufacturing Raw Materials

A valve manufacturer manufactures or purchases raw materials such as forged bonnets, seats, balls, and cast bodies. Norwegian projects are required to source raw materials from manufacturers qualified according to the NORSOK M-650 standard. In this section, we will provide more information regarding casting and forging as two of the key manufacturing processes used to produce valve parts.

6.2.2.1 Casting

Casting is a manufacturing process in which molten metal is poured into a mold, solidified, and then is removed from the mold to produce a fabricated part. According to Figure 6.1, the cast body of a pipeline valve is made of low-temperature carbon steel (LTCS), a material grade of American Society for Testing Materials (ASTM) A352LCC. Casting is a molding process that can be carried out in sand and in a ***foundry*** according to internal instructions based on a ***foundry sketch*** drawing. Figure 6.2 shows a ***basic electric furnace*** that uses an electrical arc to heat the metal during the steel-making process. The ***vacuum degassing process*** is used to enhance the cleanliness of steel after melting by removing gases such as oxygen and hydrogen. Samples are taken during the steel-making process, and chemical compositions are measured. Four integral test blocks are poured for each valve body piece (see Figure 6.3) to be subjected to mechanical testing. As mechanical tests such as Charpy impact, tensile, and hardness tests are destructive, they should be performed on test blocks.

Heat treatment is used to alter steel properties, such as mechanical strength and hardness. In order to regulate the temperature, a gas burner–heated furnace is equipped with instruments such as thermocouples. The furnace is also equipped with time and temperature recorders. The furnace should be calibrated in order to ensure that a uniform temperature is distributed throughout.

FIGURE 6.1 A pipeline valve cast body in LTCS, ASTM A352LCC. (Photo by the author.)

FIGURE 6.2 A basic electric furnace to melt the steel during casting process. (Photo by the author.)

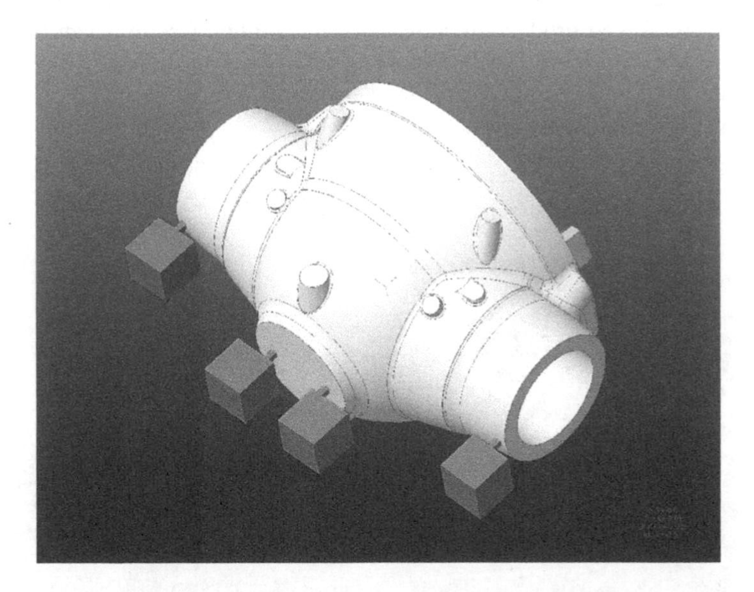

FIGURE 6.3 A pipeline valve body casting in A352 LCC with four internal test blocks. (Photo by the author.)

A variety of tests and quality control procedures are conducted on the finished casting. After the heat treatment is complete, NDT methods, such as visual inspection, magnetic particle inspection, and ultrasonic (UT) examination, are used to detect possible defects and discontinuities in the material. Known as surface NDT, visual examination and magnetic particle inspection are used on 100% of the surface area. A UT examination is a volumetric test that uses sound waves to detect defects and cracks in the components and materials being examined. The American Society of Mechanical Engineers (ASME) B16.34 standard specifies that critical areas of the valve must undergo UT testing. Manufacturers Standardization Society of the Valve and Fittings Industry (MSS) SP 55 describes the procedure for visual examination and acceptance criteria. Appendix 7 of ASME Section VIII Div.01 sets out the procedure and acceptance criteria for magnetic particle and UT examinations. During NDT, any defects found that are not acceptable are repaired by welding or grinding. Usually, weld repairs are followed by a post weld heat treatment (PWHT). After a PWHT and before a final NDT, destructive tests, including tensile, impact, and hardness testing, are performed. Following the final NDT, the material grade, manufacturer name, valve size, pressure class, pouring number, and other relevant information are marked on the casting by a low-stress stamp as shown in Figure 6.4. Chemical analysis, mechanical tests, NDT examination records, and heat treatment certificates are issued prior to casting delivery.

6.2.2.2 Forging

As a manufacturing process, forging involves shaping metal by hammering, pressing, or rolling, and it produces a higher quality product than casting. In general, the bonnet, ball, and seat of pipeline valves are forged. Cast products are close to their final shape, whereas forged products require more machining. Castings can be thought of as a process that produces parts requiring little or no machining. Similar to cast components, all forged parts require quality control, including chemical analysis, dimension control, and mechanical testing.

FIGURE 6.4 The body casting is marked with a low-stress stamp. (Photo by the author.)

6.2.3 Welding

The pipeline valves are connected to the pipeline through pups, and transition pieces have been welded to the valve's body in a welder's shop. Figure 6.5 demonstrates a 38″ Class 1500 pipeline valve's body being welded to a transition and pup piece in a welding shop, which could be performed either by the valve manufacturer or a welding subcontractor. There are four weld connections on the body of the valve pictured: two between the body and transition pieces from both sides and two between the pup and transition pieces. When the pipeline and connected valve thicknesses are less than 50% offset, the valve body is directly welded to the pup piece without any transition piece, so there are two weld joints. More information about welding technology and techniques between the pipeline valve body, transition, and pup pieces is provided in Chapter 7. A critical weld is between the lifting lugs and the valve bodies (see Figure 6.6). Alternatively, the lifting lugs can be incorporated into the valve

FIGURE 6.5 Work performed in a welding shop to join a transition and a pup piece to the body of a pipeline valve in 38″ Class 1500. (Photo by the author.)

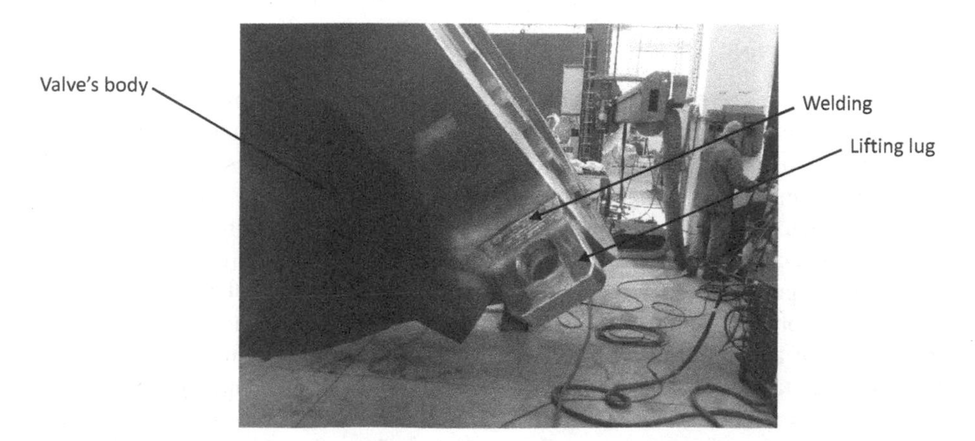

FIGURE 6.6 Welded lifting lug to the body of a pipeline valve. (Photo by the author.)

FIGURE 6.7 An LP test on the weld joint between the pup and transition piece. (Photo by the author.)

body during the casting process. A final welding step involves cladding or overlaying the Inconel 625 material on the carbon steel body or on the carbon steel grooves in the valve body.

The weld joints are subjected to a combination of NDTs, including liquid penetration (LP) and radiography (RT), to ensure that the welding operation is executed successfully without any defects. An LP test is shown in Figure 6.7 on the weld joint between the valve pup and transition pieces. The metal surface must be free from contaminants such as water, oil, and grease prior to the LP test. The test contains three liquids. The first is a red-colored penetrant which is easily drawn into defects. As for the second, a white-color developer allows for the extraction of trapped penetrant from the surface flows, and the final is a cleaner. An LP test is also required on the weld joints between the lifting lugs and the valve's body.

6.2.4 Machining

A manufacturing process called machining involves the removal of extra and unwanted materials from cast or forged products by using a cutting tool and converting them into a desired shape. Because casts are closer to the final shape than forged components, more machining is necessary for forged components. In general, there are three types of machining: ***turning***, ***milling***, and ***drilling***. Turning is a machining process in which the cutting tool is stationary and removes metal from an object that is rotating. Figure 6.8 shows the rotating forged bonnet with a still cutting tool during a turning process. Milling is the opposite of turning, where a moving cutting tool is used to cut a stationary workpiece. Figure 6.9 illustrates the milling machining process of a standing body of a pipeline valve with a rotating cutting tool. During drilling, holes are drilled into the raw materials, similar to the holes drilled in the body (see Figure 6.10) and bonnet to connect them through bolting (bolts and nuts).

FIGURE 6.8 Turning of the rotating bonnet of a pipeline valve with a fixed cutting tool. (Photo by the author.)

FIGURE 6.9 A rotating cutting tool is used to mill the stationary body of a pipeline valve. (Photo by the author.)

FIGURE 6.10 Drilled holes in the body of a pipeline valve. (Photo by the author.)

FIGURE 6.11 An LP test on the pipeline valve bonnet after machining. (Photo by the author.)

6.2.5 Inspection and Quality Control

As part of the quality control process, dimensions and NDTs (visual, LP, or magnetic particle tests) are performed following machining. The author recommends that the valve's body be subjected to a hydrostatic pressure test prior to assembly. In the event of a leakage during the test, the body is rejected and should be returned to the foundry. Figure 6.11 illustrates the LP test on the bonnet after machining.

6.2.6 Assembly of Valves

The valve's internals are positioned in the body prior to connecting the stem to the ball, and the valve's body and bonnet are then connected by bolts and nuts. The ball and seat surfaces should be lapped and hardfaced with tungsten carbide before assembly. The tungsten carbide overlay prevents wear on the ball and seat during valve operation. During the lapping process, machining errors and irregularities are removed on the ball and seat's contact areas to ensure a leak-tight joint and minimize the internal valve leakage.

6.2.6.1 Body and Bonnet Bolts Tightening

Figure 6.12 illustrates the attachment of the bonnet to the body of a pipeline valve via bolts and nuts. After the body and bonnet joints have been prepared, the correct metallic gasket is chosen and placed between the joints. Bolts and nuts are inspected and assembled to the body and bonnet; proper bolt tensioning is then applied to tighten the bolts.

Bolt torque and tension are two methods for tightening bolts. They are used to ensure that the bolts apply the required force and torque to seal a joint (e.g., tightening the body and bonnet of a pipeline valve). When a torque tool is employed, it exerts rotational force on the fasteners, while a bolt tensioning tool involves stretching the bolt before tightening the nut. Torque is defined as the amount of twisting force required to spin the nut up and along the bolt threads. Figure 6.13 illustrates a

FIGURE 6.12 Bolting the bonnet to the body. (Photo by the author.)

FIGURE 6.13 Torque tool. (Photo by the author.)

torque tool that can be used to measure torque moments. In general, torque tools are used for bolt diameters of 1″ and below, whereas tensioning tools are used for larger bolt sizes and diameters, such as 1 1/8″ and above. A total of 28 bolts, each of which is M150 equivalent to 6″ in size, are fastened to the pipeline's valve body and bonnet illustrated in Figure 6.12. The minimum and maximum torque values for tightening the bolts in Figure 6.12 are 87687 Newton meters and 131530 Newton meters, respectively. As a general rule, the minimum and maximum torque values address the torque required to tighten bolts with lubrication and cleaning and without proper lubrication. Thus, lubrication reduces friction and the amount of force and torque required to tighten bolts. The bolt force or tension to achieve an acceptable torque value is 7200 kN (kilo-newtons), and the maximum bolt tension before plastic deformation is 9238 kN. As a torque tool cannot provide sufficient force or torque to tighten the bolts in the picture, bolt tensioning devices must be used. Bolt tensioning tools are typically operated by hydraulic force. Bolt torque is determined by three factors: bolt size, bolt material, and lubrication of the bolt and nut. Generally, larger bolt diameters, stronger materials, and less effective lubrication require greater torque for tightening. Figure 6.14 illustrates the process of tightening the bolts and nuts belonging to the pipeline valve's body and bonnet, using four bolt tensioning tools. As a rule of thumb, the maximum bolt tension or force supplied by the torque or tensioning tool should be at least equal to the acceptable bolt force which is 7200 kN. For example, a bolt tensioning tool that produces 8505 kN through a hydraulic oil pressure of 150 bar is sufficient.

For bolts larger than 1″, such as the ones in this case, hydraulic bolt tensioning is the preferred method of tightening them. When a hydraulic bolt tensioning method is employed, the bolt length should be increased by a value equal to the bolt diameter. With the additional length of the bolt, there are enough bolt threads to engage inside the hydraulic bolt tensioning tool. Hydraulic bolt tensioners are hollow, hydraulic, compact cylinders with internal threads that engage with the threads of bolts. The

FIGURE 6.14 Bolt tensioning tools are used to tighten bolts and nuts on a pipeline valve's body and bonnet. (Photo by the author.)

hydraulic oil is supplied by applying pressure or load in order to pull out the stud bolt. Bolt tensioning is achieved by pulling out the bolt and applying direct axial force with a tensioning tool. In applying the axial load under hydraulic pressure, the stud bolt is stretched, creating compression on the joint at the same time. The axial force applied on the bolt is transferred to the nut, and the nut moves upward with the bolt. Therefore, the operator turns down the nut under the tensioning device to fasten the joint. Once the nut has been tightened, the hydraulic pressure can be released, and the tensioning device can be removed.

The bolts are tightened by following a method called crisscrossing or simply "crossing." A typical crisscross bolt tightening sequence can be seen in Figures 6.15

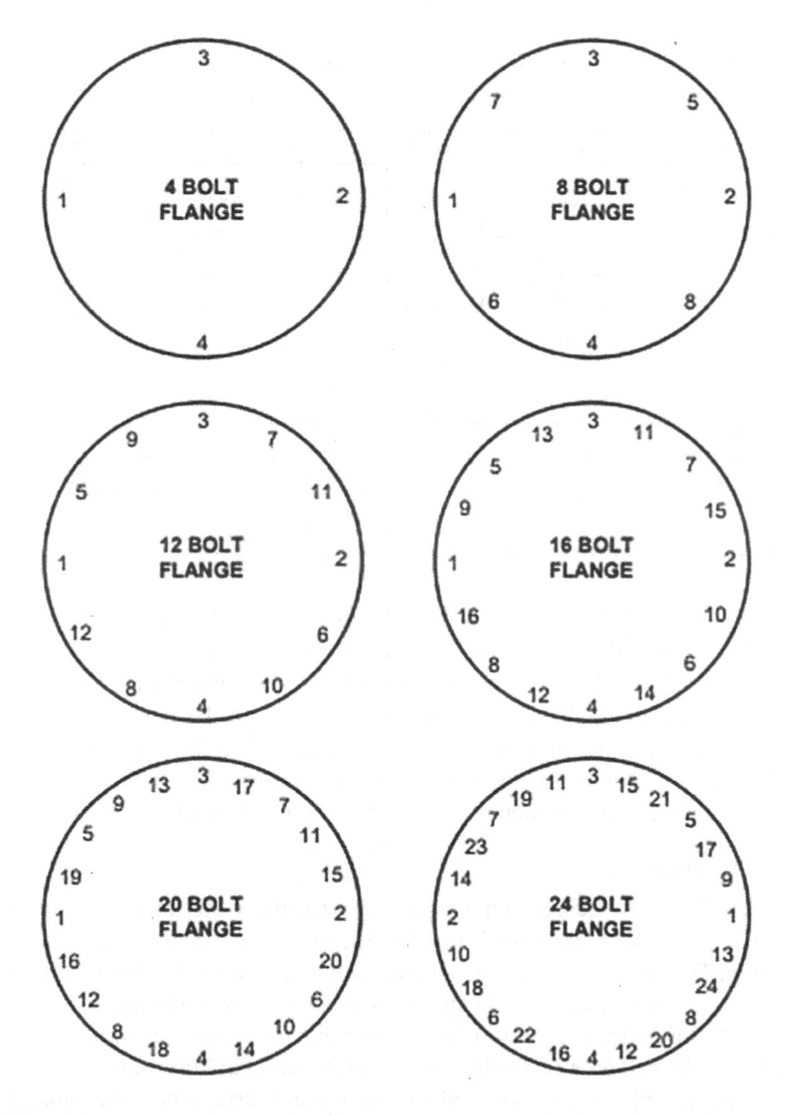

FIGURE 6.15 Typical crisscross bolt tightening sequence for joints with 4, 8, 12, 16, 20, and 24 bolts.

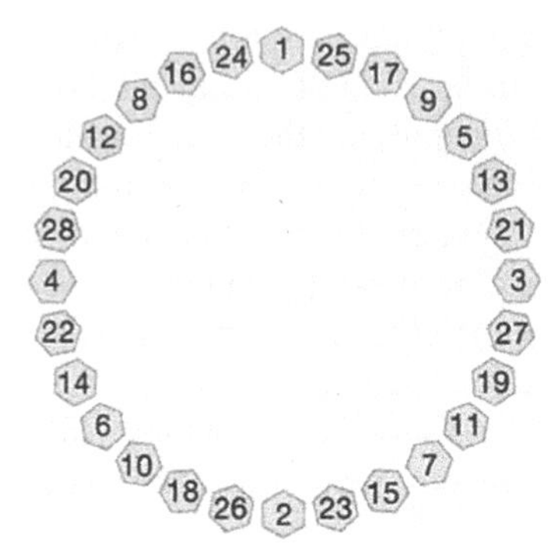

FIGURE 6.16 Typical crisscross bolt tightening sequence for a joint with 28 bolts.

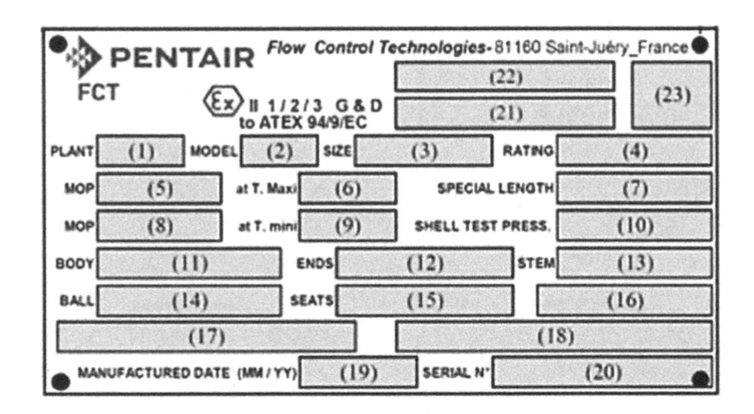

FIGURE 6.17 Nameplate of a pipeline valve.

and 6.16 for joints with 4, 8, 12, 16, 20, 24, and 28 bolts. It is critical to follow the numbered sequence of the bolt holes on the flanges when fastening the bolts. Bolts can be numbered on the flanges to simplify the process of tightening the bolts. As soon as the torque tool indicates that the correct torque has been achieved by "clicking" or "breaking," it is important to stop overtightening the bolt. Otherwise, the over-torque may result in overstressing the joint, resulting in cracking, gasket damage, and other undesirable outcomes resulting in failure of the body and bonnet joints.

6.2.6.2 Marking

In accordance with the requirements of the project, the valve manufacturer attaches a nameplate or tag plate (see Figure 6.17) to the valve body containing the necessary information. The top of the tag plate includes the name of the valve manufacturer (e.g., Flow Control Technologies or FCT) and the address (e.g., 81160 Saint-Juéry, France). The ***Ex marking*** indicates explosion protection and a product's compliance with the ***Atmosphere EXplosible (ATEX)*** directives. ***ATEX*** is a set of European regulations that are aimed at ensuring the safety of products such as industrial valves in explosive environments. On the rectangular space (1), the name of a project or plant is written. On the right side of the plant designation is the valve model, size, and rating. There is also important information on the nameplate regarding the operating conditions, including both pressure and temperature. (7) and (10), respectively, describe the end-to-end measurement of the valve as well as the test pressure for the valve shell. Test pressure values and related calculations are discussed in Chapter 9. On the nameplate, materials of essential components such as the body, ball, seats, and stem are highlighted. The valve serial number and manufacturing date are located at the bottom of the nameplate.

6.2.7 Testing

Valve testing, including various types of testing such as hydrostatic and gas pressure tests and function tests described in Chapter 9 in detail, is performed after the assembly has been completed. All valves are internally tested prior to the witness test when a third-party inspector or a representative of the client (purchaser) is invited to the ***factory acceptance test (FAT)***. Testing pipeline valves after their design, manufacture, and assembly is essential for identifying potential valve defects. Valves require testing to ensure that they operate correctly and accurately without any failures or leaks during their operational lifetime, which may be as long as 20–30 years. When a valve fails due to insufficient testing or for any other reason, it must be maintained or replaced. The maintenance and replacement of valves, especially pipeline valves, during operation is extremely costly. This is because it can result in process or plant shutdown, loss of assets, and the replacement of a part or the entire valve. Moreover, valve failure and the leakage of a hazardous, highly pressurized, and potentially flammable fluid outside the valve pose significant health, safety, and environmental risks such as environmental pollution and harm to onsite personnel. Figure 6.18 illustrates the process of testing an electrically actuated pipeline valve in a pit and the benefits of such testing.

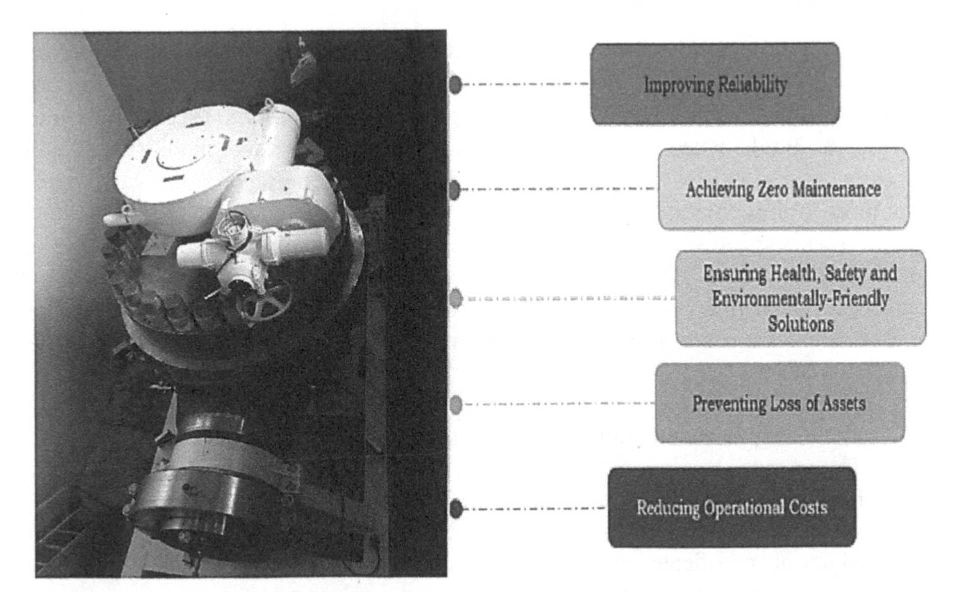

FIGURE 6.18 Testing of an electrically actuated pipeline valve (left) and the benefits of the testing (right). (Photo by the author.)

6.2.8 Coating or Painting

Pipeline valves in the offshore industry are typically coated. The main purpose of coating is to prevent external corrosion caused by the harsh and corrosive offshore environment. If the external surface of the valves is painted or coated during the shell or body pressure test, this can conceal possible leakage from the valve. Figure 6.18 illustrates a pipeline valve during the pressure test, which is not coated. It should be noted that an internal leakage test can be conducted if the valve has a coating.

6.2.9 Preservation and Packing

A valve, including pipeline valves, must be preserved during and after delivery for a specific period of time (e.g., two years) as specified in the project documents, such as the valve specification, to protect it from adverse environmental conditions. Wind, dust, humidity, salt, sandblasting, painting, and high or low environmental temperatures are harmful environmental conditions. Valve protection involves all stages of transportation, storage, lifting, handling, installation, and testing in the construction yard. Duration of preservation depends on the storage and fabrication period, ranging from six months as a minimum to two to five years as a maximum. To prevent damage during handling, transportation, and storage, all equipment and facilities, including valves, should be packed or appropriately boxed. In addition, the packing should protect the valves from dust, humidity, and external mechanical forces that may occur during storage, loading, or unloading. Additional information on packing and preservation is provided in Chapter 10.

6.2.10 Final Inspection

It is the client or a third-party inspector who performs the final inspection before shipment. The final inspection includes checks regarding preservation, dimensions, marking and tagging, and packaging. Valves experts recommend that during the final inspection all information on general arrangements (GAs) and cross-section drawings be checked. A GA drawing is generated by the valve manufacturer to demonstrate the external parts and key dimensions of the valve. The valve cross-section drawing is not dimensional, and it provides the description and material of each valve component. The valve manufacturer usually invites an inspector for a final inspection two or three weeks after implementing pressure testing. Here are some typical valve problems that are often discovered during the final inspection:

- The cavity vent and drain blind flanges have the wrong surface finish
- The proper cleaning and drainage of the water after the pressure test have been forgotten (see the case study in Chapter 10)
- An incorrect or poorly marked valve (e.g., the markings are difficult to read)
- Materials certificates are not in accordance with valve materials
- Incorrect coating systems or thickness on the valves and actuators

6.2.11 Shipment

The assembled valves and actuators are shipped to the client with the packing list, invoices, and other documents required for installation in the construction yard. Transportation of pipeline valves is explained in detail in Chapter 11.

QUESTIONS & ANSWERS

1. What document provides an overview of industrial valves–related manufacturing, testing, and inspection activities?

   A. Datasheets
   B. Inspection and test plan (ITP)
   C. Packing and preservation procedure
   D. Material take off (MTO)

   Answer) Datasheets provide detailed technical information about the valves, such as materials and standards. Thus, option A is incorrect. The correct response is option B. The packing and preservation procedure is a document prepared by the valve supplier to address only packing and preservation issues. Therefore, option C is incorrect. Option D is incorrect since the MTO is an engineering document which lists the valves and their quantities.

2. Select the wrong sentences about the casting manufacturing process.

   A. By casting a valve component, a product is produced that is close to the desired shape and requires less machining than forged components.
   B. To perform non-destructive testing (NDT) on valve body parts, integral test blocks are poured.
   C. No welding repairs are permitted on the casting.
   D. After the final NDT, the material grade, manufacturer name, valve size, pressure class, pouring number, and other relevant information are stamped on the casting using a low-stress stamp.

   Answer) Options A and D are correct statements. Option B is incorrect, as the main reason for integral test blocks is to perform mechanical tests on them and not NDTs. Option C is not correct either because weld repair is allowed on cast pipeline valve bodies. Thus, options B and C are wrong.

3. What is the condition in which an NDT is required?

   A. After the valves' components have been machining
   B. After welding the pup piece to the valve's body
   C. After welding lifting lugs to the body of the valve
   D. All options are correct

   Answer) In the valve industry, NDT is applied after machining and welding to detect possible defects in the machined surfaces and weld joints. Therefore, option D is the correct answer.

4. Which of the following is not a benefit of factory acceptance test?

   A. Reliability improvement
   B. Loss of asset prevention
   C. Health, safety, and environment (HSE) improvement
   D. Reducing the initial or capital cost of the valve

   Answer) Option D is the correct answer. Reducing the initial or capital cost is not a consequence or benefit of FAT.

5. Which sentence is correct?

   A. The valves are pressure tested after they have been coated
   B. Cast-made parts are of better quality and more expensive than forged components
   C. Turning is a machining process in which the cutting tool does not move and removes the metal from a rotating work piece
   D. The purpose of preservation is to protect valves only during manufacturing

   Answer) Option A is incorrect since the valves are pressure tested without coating. Option B is also incorrect since forging components are both more expensive and of higher quality than casting components. The correct response is option C. As preservation is used during manufacturing, handling, transportation, storage, and installation, option D is incorrect. In summary, option C is the correct option.

MORE PICTURES

Figures 6.19–6.22 show some stages of manufacturing, assembly and shipment of a large pipeline value.

FIGURE 6.19 A pipeline valve stem during the machining operation. (Photo by the author.)

FIGURE 6.20 An LP test was carried out on the welding between the valve body and the lifting lug. (Photo by the author.)

FIGURE 6.21 Lifting a ball during valve assembly in order to place it inside the body during valve assembly. (Photo by the author.)

FIGURE 6.22 Shipment of a pipeline valve. (Photo by the author.)

BIBLIOGRAPHY

1. American Society of Mechanical Engineers (ASME) B16.34. (2017). *Valves–Flanged, threaded, and welding end*. New York, NY: ASME.
2. American Society of Mechanical Engineers (ASME). (2019). Rules for construction of pressure vessels. Boiler and Pressure Vessel Code. ASME Section VIII Div.01. New York, NY.
3. Condor Machinery. (2021). What, exactly, is machining? Popular types, tools and techniques. [online] available at: https://www.condormachinery.com/what-exactly-is-machining-types-tools-techniques/ [access date: 26th December 2021].
4. Engineering articles. (2021). Machining operation and types of machining tools. [online] available at: https://www.engineeringarticles.org/machining-operation-and-types-of-machining-tools/ [access date: 26th December 2021].
5. Mades N. (2014). Understanding about inspection and test plan (ITP). [online] available at: https://www.qualityengineersguide.com/understanding-about-inspection-and-test-plan-itp [access date: 25th December 2021].
6. Manufacturers Standardization Society of the Valve and Fittings Industry (MSS) SP 55. (2011). Quality standard for steel castings for valves, flanges, fittings, and other piping components- Visual method for evaluation of surface irregularities.
7. NORSOK M-650. (2011). *Qualification of manufacturers of special materials*, (4th revision). Lysaker, Norway.
8. Sotoodeh K. (2015). Top entry export line valves design considerations. *Valve World Magazine*, 20(05), 55–61.
9. Sotoodeh K. (2016). Valve Preservation Requirements. *Valve World Magazine*, 21(05), 53–56.

10. Sotoodeh K. (2019). Managing valves in EPMS projects. Valve magazine. [online] available at: https://www.valvemagazine.com/articles/managing-valves-in-epcm-projects [access date: 25th December 25, 2021].
11. Sotoodeh K. (2021). A practical guide to piping and valves for the oil and gas industry, (1st edition.) Austin, USA: Elsevier (Gulf Professional Publishing).
12. Sotoodeh K. (2022). *Piping engineering: preventing fugitive emission in the oil and gas industry*, (1st edition). New York: Wiley.
13. Sotoodeh K. (2022). *Cryogenic valves for liquified natural gas plants*. (1st edition). Cambridge: Elsevier (Gulf Professional Publishing).

7 Welding Technology

7.1 INTRODUCTION

A wide variety of end connections are available between valves and piping, including flanges, threaded connections, mechanical joints, and welding. In the first part of the chapter, various types of end connections are described. Butt welding is the most appropriate joint for connecting offshore pipeline valves to the connected pipeline. Furthermore, this chapter discusses two main welding types; the first pertains to the connection of the transition and pup pieces to the valve body, as well as the connection of the valve to the pipeline. In addition, Inconel 625 is used as a weld overlay or cladding in the grooves of the body and bonnet beneath the location of seals and gaskets in order to prevent crevice corrosion. We will discuss these end connections and their advantages and disadvantages in the following section, as well as why pipeline valves are welded to the pipeline.

7.2 VALVE END CONNECTIONS

7.2.1 Standard ASME Flange

Flanged joints consist of two mating flanges connected by bolts and nuts, and a gasket is positioned between the flanges to prevent leakage. A major advantage of flanged end valves is that they allow access to the valve's internal components for cleaning and maintenance purposes. However, using flanged ends for pipeline valves, as discussed in Chapter 1, makes them and the connected pipeline systems heavier and increases the possibility of leakages. In Figure 7.1, there is an illustration of a 20″ ball valve in pressure class 150 (pressure nominal of 20 bar) with flanged end connections. Flanges and flanged ends integrated into valve bodies are typically made to American Society of Mechanical Engineers (ASME) relevant standards, including ASME B16.5 and ASME B16.47.

7.2.2 Mechanical Joint (Hubs and Clamps)

The hub and clamp connection is another type of connection used as an alternative to ASME flanges to save space, weight, and cost. Mechanical joints consist of two hubs, two clamps, one seal ring, four stud bolts, and eight nuts. Hub and clamp connections are widely used in the Norwegian offshore industry for high-pressure classes, such as 1500 and 2500 ASME standards, equivalent to 250 and 420 psi, respectively, and sizes above 2″. Figure 7.2 shows a ball valve in the high-pressure class of 1500 with blind hub ends, and the valve has been prepared for testing by blinding the right end of the valve with a blind hub and clamps.

DOI: 10.1201/9781003343318-7

FIGURE 7.1 A 20″ ball valve in pressure class 150 with two flanged end connections. (Photo by the author.)

FIGURE 7.2 A high-pressure class hub end ball valve. (Photo by the author.)

7.2.3 Compact Flange

The NORSOK L-005 compact flange series provides an alternative to ASME flanges for high-pressure classes (e.g., class 600 equal to 100 bar and higher) and pipe sizes

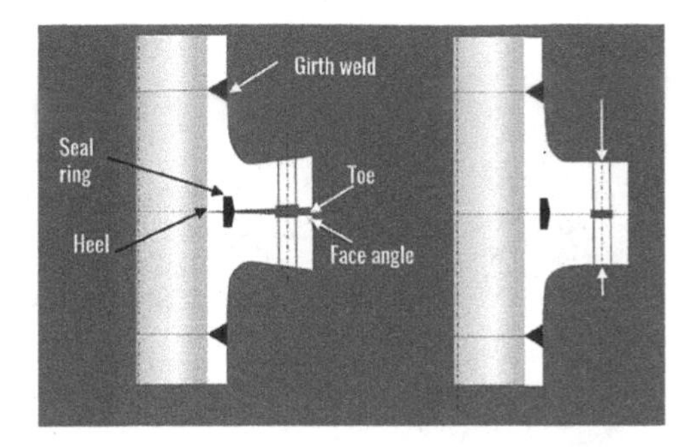

FIGURE 7.3 Compact flange double sealing faces of heel and ring. (Photo by the author.)

of 3″ and larger. In contrast to ASME flanges, compact flanges are not as compact as mechanical joints, but they also offer the following advantages:

- More resistance to stresses and loads
- Savings in space due to smaller flange dimensions
- Provides more safety and fewer leakage possibilities due to the two sealing faces on the heel and ring on the left side of Figure 7.3.
- Because the gasket in compact flanges is less likely to suffer corrosion due to no contact with internal fluid and external corrosive offshore atmosphere, this results in greater reliability and fewer leakage risks.

There is one major disadvantage of the compact flange, and that is that gaskets are vulnerable to damage on their faces. As a result, compact flanges are not recommended for rotary equipment (e.g., pumps and compressors) and piping joints with spectacle blinds, spades, and spacers. The purpose of spades in piping systems is to provide pressure retention between two flanges in order to isolate or shut down a portion of the pipeline in order to interrupt the flow of liquids temporarily or permanently for purposes such as leak detection and maintenance. A spacer serves the same purpose as a spade in a piping system to maintain the pressure between two flanges. Figure 7.4 illustrates the installation of a typical spade, spacer, and spectacle blind between flanges. Basically, a spacer is used when the pipe needs to be opened, and a spade is used when the pipe needs to be blocked. To switch between a spade and spacer, the flange bolts should be loosened. As a spectacle blind has both a spade and a spacer built into it, it can be used easily to switch between the two. The face of the compact flanges can be damaged by both rotating equipment applied loads and frequent switching between a spade and spacer.

7.2.4 Wafer Connection

A wafer is defined as a flangeless design with a facing that permits installation between ASME and manufacturer standard (MSS SP) flanges. With a wafer design, the valve's face-to-face space, weight, and cost are reduced as compared to a flanged end design. A wafer design allows long-length bolts to be passed along the valve body and through two flanges on both sides of the valve, as shown on the right side

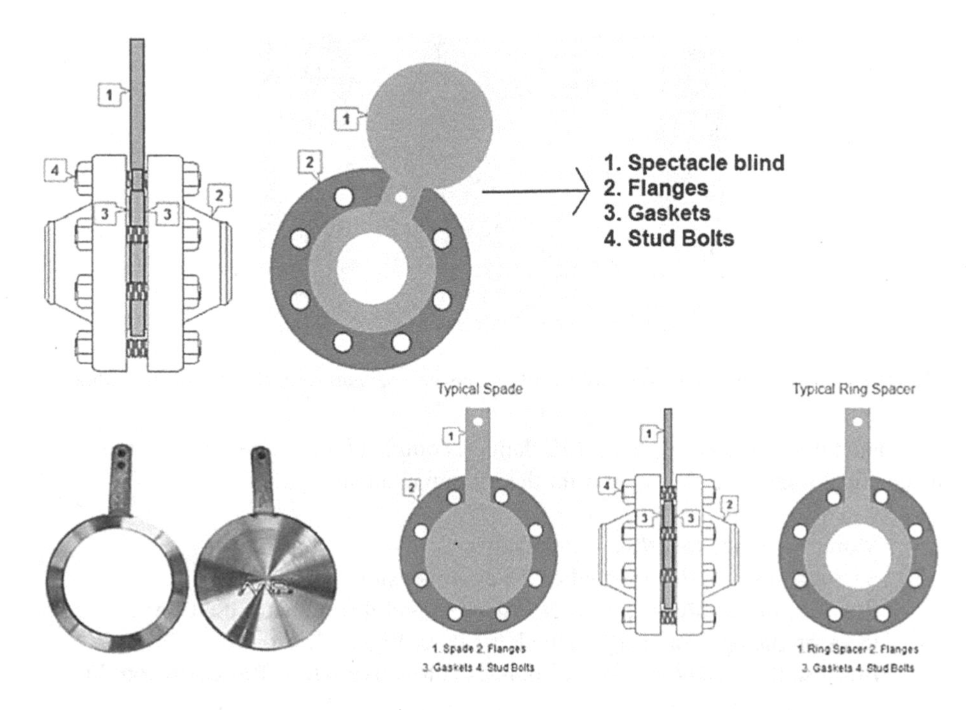

FIGURE 7.4 Spade, spacer, and spectacle blinds and their installation between flanges. (Courtesy: Facebook, Inspection academy.)

of Figure 7.5. Lugged or wafer lug connections are similar to wafer type connections, with two exceptions: the lugged holes are threaded, and the threaded lug holes accept bolts from both sides. Unlike wafer lug connections, lugged connections (see Figure 7.5 left side) can be used for the end of a line where the flange at one end can be opened toward the atmosphere, and the lugged valve can be held in place by the bolts at the other end. Check valves and butterfly valves can be selected as lug or wafer connections; however, lug or wafer connections are not applicable for pipeline valves because they are not robust.

7.2.5 Threaded Connection

This type of pipe joint is typically used for small pipe diameters in nominal pipe sizes (NPS) of 2″ or less. The thread end piping consists of a male part that is inserted into a female part of the valve or fitting. The ASME B 1.20.1 standard contains the dimensional requirements for taper pipe threads. There is a risk of leakage through the threaded connection due to temperature change, vibration, and high stress concentration, as well as an increase in corrosion. It is used in low-cost, non-critical applications such as water in onshore facilities, such as refineries and petrochemical plants. In contrast, threaded piping connections are not used in the subsea and topside offshore sectors of the oil and gas industry. This is because the failure of facilities, including piping, is less tolerated in the offshore sector. To summarize, the threaded connection does not apply to pipeline valves since they are large and subject to large forces.

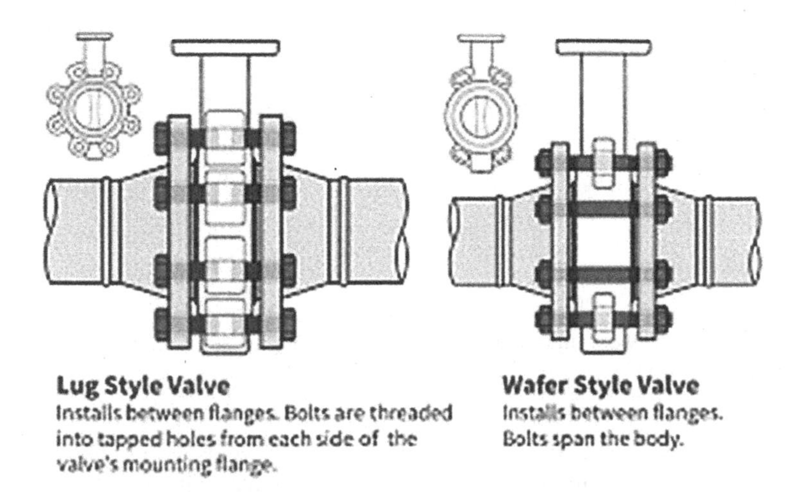

FIGURE 7.5 Wafer and lug style valves as per API 609 standard.

7.2.6 Welded Connections

Welded connections are joined together through welding, a process in which the connected joints are heated and fused while the extra metal is supplied and deposited from an electrode. In this section, we will discuss two types of welding: a socket weld and a butt weld.

7.2.6.1 Socket Weld

Contrary to a butt weld connection, a socket weld connection can be used for small pipe connections up to 2″ in diameter. A socket weld pipe is the male part inserted into a recessed area or into the female part of a valve or fitting. Fittings or valves suitable for socket welding connections to pipe up to the size of 4″ comply with ASME B16.11. It is imperative to note, however, that although socket welding fittings are defined in the standard for a maximum size of 4″, it is very common to use them for a maximum size of 2″. The type of socket weld shown in Figure 7.6 between the pipe and valve body or fitting is a fillet weld.

When assembling a socket weld joint, the male end of the pipe should be inserted into the socket to the maximum depth and then withdrawn approximately 1/16″ (1.6mm) inward from the shoulder of the socket (refer to the ASME B31.3 code). The gap created is most likely intended to cover the expansion of the inserted male part as a result of heat produced during welding. As a result of the expansion gap, a crevice forms within the socket welding joint, which can promote crevice corrosion. As a result, the present author does not recommend socket welding for corrosive applications such as seawater. In contrast, piping engineers can use socket weld piping for handling non-corrosive process fluids. A non-destructive test (NDT) may be applied to socket welds without destroying them to ensure that the joints are free of defects and discontinuities. The other weakness of the socket weld is that it has less strength compared to a butt weld connection. In summary, the main reason that socket weld connections are not used for pipeline valves is that these valves are

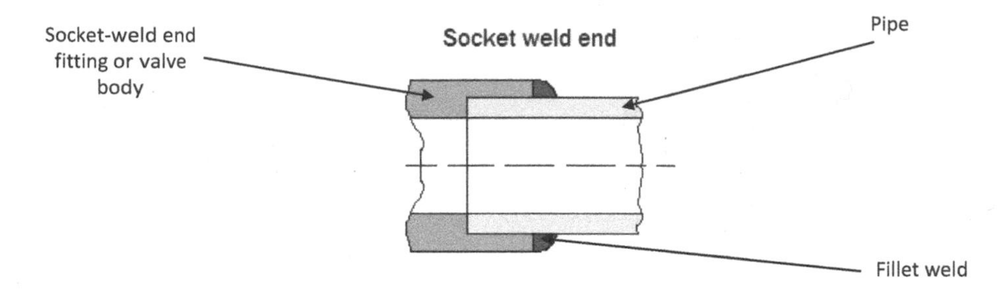

FIGURE 7.6 Socket weld (fillet weld) between a piece of pipe and a valve or fitting. (Photo by the author.)

significantly larger than 2″ in diameter, whereas socket welds are limited to a maximum 2″ diameter. In addition, the risk of leakage is greater with a socket weld than with a butt weld. However, the possibility of leakage from a socket weld is significantly lower than that from a threaded joint.

7.2.6.2 Butt Weld

Welders perform a butt weld connection between two joints with ***bevel ends*** (BE). This type of pipe ending, BE, has a lean angle with the pipe surface and is designed for butt weld connections. As illustrated in Figure 7.7, the pipe end for a butt weld joint is typically prepared in accordance with ASME B16.25. Near the welding root, a ***root face*** of approximately 1.6 millimeters is prepared in order to prevent the collapse of the welding. Figure 7.7 illustrates two different butt weld end connection arrangements. The design on the top is applicable to connections with a maximum thickness of 22 mm, and where the pipe connections' ends are normally chamfered at a 37.5-degree angle. The butt weld end preparation for piping connections more than 22 mm thick requires two chamfers. The first one is 37.5° until 19 mm thickness and the second chamfer is 10° on average.

In addition, narrow gap welding may be applied to butt weld end connections that include a chamfer at a smaller degree than 5–7°; these can be welded more quickly and with fewer electrodes (welding consumables). However, although some concerns have been raised regarding the lack of fusion in narrow gap welding, this type of butt weld preparation and connection has been used successfully in the oil and gas industry. In a recent Norwegian offshore project, Figure 7.8 depicts a narrow gap welding performed between two piping components during the fabrication of a topside manifold constructed out of 22 Chromium (Cr) duplex material. The figure on the left shows two piping components after fitting up, and the figure on the right shows the welded joints, along with the deposited welding consumables. The manifold is an arrangement of piping used on platforms or ships to gather the produced fluid or gas from different wellheads. It also allows operators to inject water or gas into the wellhead as part of an advanced oil recovery process. Manifolds are primarily designed to reduce the number of piping connections by combining them.

In ASME B16.25, the butt weld ends of piping components connected to the piping system are prepared. The butt weld end connection is the most commonly used

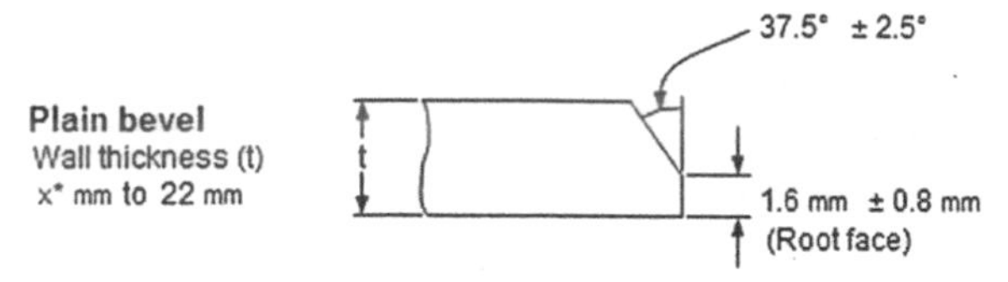

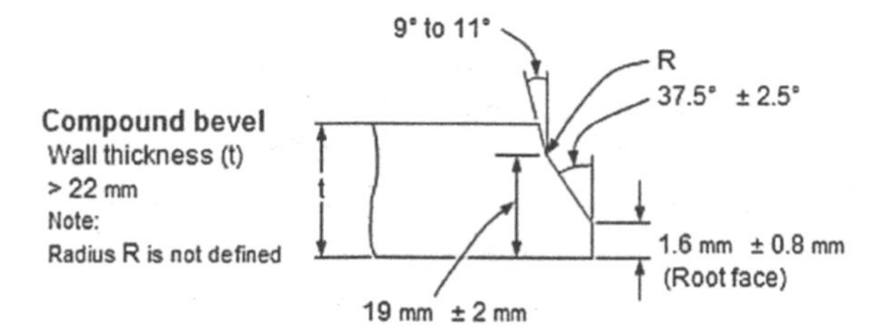

FIGURE 7.7 Bevel end preparation as per ASME B16.25.

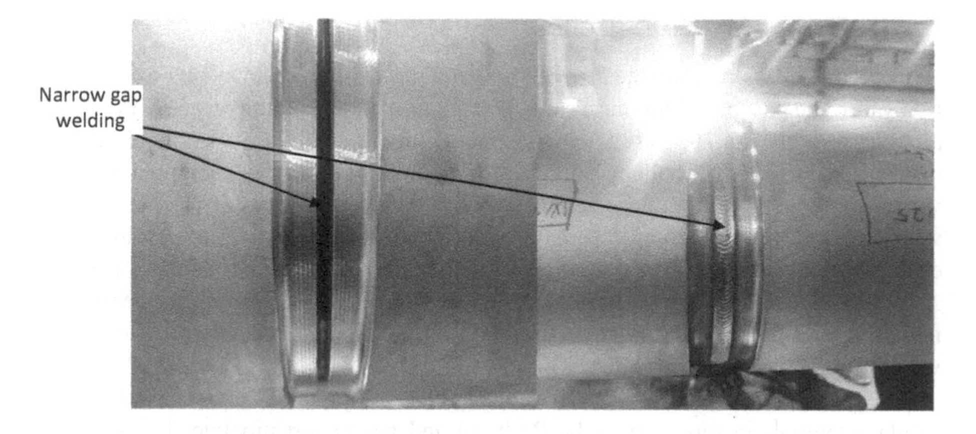

FIGURE 7.8 A narrow gap welding technique was employed during the fabrication of the manifold in order to connect two piping components in a 22Cr duplex. (Photo by the author.)

welding connection in the oil and gas industry, also known as ***circumferential welding***. The use of a butt weld connection between the valve and pipe rather than a flange connection provides higher ***joint efficiency*** and lower leakage risks. According to various piping codes (e.g., ASME B31.3, process piping code) and standards as well as other literature, joint efficiency is defined as the ratio of the welded part strength to the strength of the base material (see Equation 7.1).

Equation 7.1 Weld joint efficiency or joint factor calculation

$$E = \frac{S_w}{S_b}$$

where:

E: weld joint efficiency or factor;
S_w: weld strength (Ksi, Pascal);
S_b: base metal strength (Ksi, Pascal).

The joint efficiency for a butt weld joint could be between 0.8 (80%) to 1 (100%) depending on the type of butt weld and whether a 100% radiography test was applied on the joint. Using 100% radiography examination on the connected joints increases the welding joint efficiency to 100% and reduces the risk of leakage from the connection to zero. The leak-proof and strength of butt weld joints make this type of connection suitable for pipeline valves. The next section focuses on a case study addressing the butt weld connections between oil and gas export pipeline valves and a couple of pipelines in an offshore oil and gas project.

7.3 A CASE STUDY OF WELDING BETWEEN PIPELINE VALVES AND TWO OIL AND GAS EXPORT PIPELINES

7.3.1 Pipeline Material Selection

Workers on site weld between valves and oil and gas export pipelines in the construction yard. The purpose of this case study is to examine the welding of a 38″ oil export pipeline as well as a 20″ gas export pipeline to pipeline valves. It is critical to note that one of the primary challenges in welding operations is the fact that pipelines and valves are made of dissimilar materials; pipelines are made from 22Cr duplex, while valves are made from carbon steel. After pipeline inspection gadgets (pig) launchers, each pipeline has three valves. In a pig launcher, a pressurized container is used to shoot a device called a "pig" through a pipeline to perform a variety of operations, including cleaning, monitoring, and maintaining the pipeline. Chapters 1 and 3 provide more information on pigging a pipe.

This paragraph explains why valves and connected pipelines are selected in dissimilar materials in this case study. Both oil and gas export pipelines can be constructed from non-corrosion-resistant alloys, such as carbon or low-alloy steels, since they transport well-treated and non-corrosive petroleum products. However, both pipelines are constructed of 22Cr duplex material in order to reduce weight. In Chapter 4, we discussed the importance of weight reduction for equipment located above offshore platforms. What are the benefits of upgrading the material of the pipeline from carbon steel to 22Cr duplex? As the wall thickness of a pipe or pipeline increases, it becomes heavier. The weight of a pipe can be calculated using Equation 7.2.

Equation 7.2 Calculation of a pipe or pipeline mass weight

$$m = \rho * V = \rho * l * A = \rho * l * \frac{\pi}{4}\left(OD^2 - ID^2\right) = \rho * l * \frac{\pi}{4}\left(OD^2 - (OD - 2t)^2\right)$$

where:

m: pipe or pipeline weight (kg, pound);

ρ: pipe or pipeline density $\left(\frac{Kg}{m^3}, \frac{pound}{inch^3}\right)$;

V: pipe or pipeline volume (m^3, $inch^3$);

l: pipe or pipeline length (m, inch);

OD: pipe or pipeline outside diameter (m, inch);

ID: pipe or pipeline inside diameter (m, inch);

t: pipe or pipeline wall thickness (m, inch).

Using the equation, the outside diameter of a pipe or pipeline, parameter OD, is a fixed value that depends only on the pipe size. It does not change by changing the wall thickness. In contrast, the internal diameter (ID) of a pipe or pipeline is reduced by increasing the wall thickness value. The reduction of ID according to Equation 7.2 leads to an increase in pipe or pipeline weight. A variety of ASME codes provide pipe or pipeline wall thickness calculation methods. These include ASME B31.3 process piping, ASME B31.4 pipeline transportation systems for liquids and slurries, and ASME B31.8 gas transmission and distribution piping systems. Often, the ASME B31.4 code is used to design pipelines containing liquids, such as oil export pipelines, including calculations of wall thickness. Since fewer hazards are involved with liquid pipelines, the ASME code for liquid-containing pipelines is less conservative than ASME B31.3, the process piping code. A pipeline failure caused by internal pressure, however, should be taken seriously, as this could lead to emissions, major failures, and the loss of assets, etc. The method for calculating the wall thickness of a gas pipeline as per ASME B31.8 code is less stringent than that of ASME B31.3 code, but more stringent than ASME B31.4 code. In order to ensure a higher level of safety, the design code for oil and gas pipelines, including their wall thickness calculations, may be based on ASME B31.3, rather than ASME B31.4 and B31.8. Equations 7.3, 7.4, and 7.5 are used to calculate pipe or pipeline wall thickness calculations using the codes mentioned previously.

Equation 7.3 Piping wall thickness calculation as per ASME B31.3 code

$$t = \frac{PD}{2(PY + SE)}$$

where:

t: pipe wall thickness (inch);

P: design pressure (psi);

D: pipe outside diameter (inch);

Y: 0.4 (material coefficient);

S: allowable stress (psi);

E: joint efficiency (dimensionless).

Equation 7.4 Pipeline wall thickness calculation as per ASME B31.4 code

$$t = \frac{PD}{2\,FES_Y}$$

where:

t: minimum design wall thickness (inch);
P: internal pressure in the pipe (psi);
D: outside diameter of the pipe (inch);
F: design factor equal to 0.72 throughout;
E: longitudinal weld joint factor, usually a value from 0.8 to 1;
S_Y: minimum yield stress for the pipeline (psi)

Equation 7.5 Pipeline wall thickness calculation as per ASME B31.8 code

$$t = \frac{PD}{2\,FETS_Y}$$

where:

t: minimum design wall thickness (inch);
P: internal pressure in the pipe (psi);
D: outside diameter of the pipeline (inch);
F: design factor, which is dependent on location (dimensionless);
E: longitudinal weld joint factor, usually a value from 0.8 to 1 (dimensionless);
T: temperature derating factor (dimensionless);
S_Y: minimum yield stress for the pipeline (psi)

For all three equations, the pipe wall thickness is determined by the mechanical properties (e.g., yield strength or allowable stress) of the materials used for piping or pipelines. The yield strength of a material is the point at which the material begins to deform plastically. This means that it deforms permanently and cannot be returned to its original state beyond the yield strength. The allowable stress is less than the yield strength and is defined as the maximum stress that can be safely applied to a component, equipment, or structure. The amount of thickness is inversely proportional to the yield strength and allowable stress values. As these two mechanical properties increase, the wall thickness of the pipeline or piping will decrease. Since 22Cr duplex material has higher mechanical properties, including yield and allowable stress, than carbon steel, changing the material of a pipe or pipeline from 22Cr duplex to carbon steel results in a reduction in wall thickness. In addition, extra millimeters (e.g., 3 mm as per NORSOK standard) are added to the wall thickness of piping and valves in non-corrosion-resistant alloys (CRAs), such as carbon and low-alloy steels, as a corrosion allowance to mitigate the severity and high risk of carbon dioxide corrosion. In contrast, 22Cr duplex is a CRA material that is not subject to carbon dioxide corrosion. As a result, there is no need to increase the thickness of a pipe or valve in 22Cr duplex. Thus, due to its higher mechanical properties and superior corrosion resistance against carbon dioxide, 22Cr duplex pipes are lighter than carbon steel pipes of the same size and pressure class. The offshore pipelines used to transport oil and gas from the platform to the shore in this case study are divided into two sections. The first part is situated on a platform and is made of 22Cr duplex material in accordance with ASME standards for the reasons cited above. In the second section, there is a subsea pipeline that is designed in accordance with the standards of Det Norske Veritas (DNV) and is made of low-alloy steel API 5L X65. API 5L specifies API 5LX grades, which are high-strength low-alloy steel pipes such as X42, X46, X52, X56, X60, X65, X70, and X80. The two-digit number after the "X" indicates the yield strength of the grade in ksi (kilo pounds per square inch). For example, X65 pipe has a minimum yield strength of 65 ksi.

Figure 7.9 illustrates a schematic of three pipeline valves on an oil export pipeline in this case study, including four highlighted sections of pipeline. The first pipeline section is called "Part 1," and it is located between the pig launcher and the first oil export pipeline valve, which is situated above the platform and is made of 22Cr duplex steel. Second pipeline section, referred to as "Part 2," is situated between the first and second pipeline valves and is constructed of 22Cr duplex material. "Part 3," as well as the previous two parts, is made of 22Cr duplex and is located between the second and third pipeline valves. Part 4 is the subsea pipeline fabricated from API X65 material, and it connects to the third pipeline valve from the left side. The code break between DNV and ASME occurs where the third pipeline valve is welded to the subsea pipeline. While the material used in pipelines is upgraded to 22Cr duplex to save weight, the bodies and bonnets of pipeline valves are made from non-CRAs such as carbon or low-temperature carbon steel (LTCS) since such a change for pipeline valves would be quite costly. In the event the valve's body is made from 22Cr duplex, all its internals, called trim, should be constructed from 22Cr duplex or a material with higher corrosion resistance than a duplex.

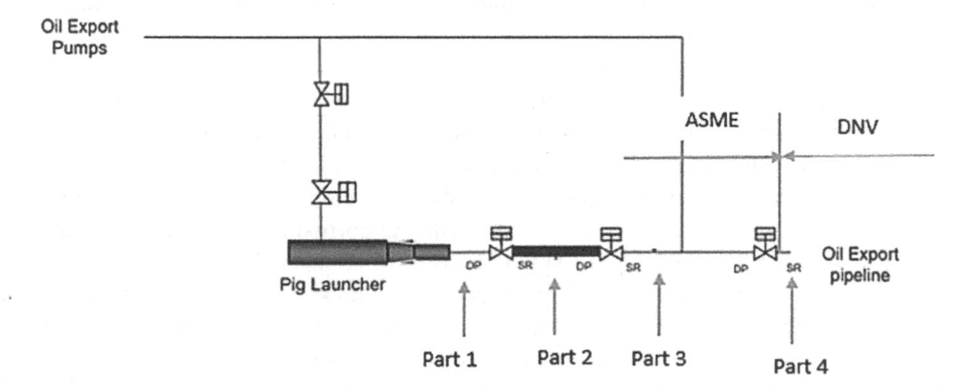

FIGURE 7.9 Schematic representation of three pipeline valves on an oil export pipeline in the case study. (Photo by the author.)

7.3.2 Pup and Transition Pieces

A valve manufacturer or a welding subcontractor should weld each valve body to at least one piece of forged pipe (e.g., pup or transition piece) on each side. Valve circumferential welds between the valve body, pup, and transition pieces are performed before valve assembly and placing the valve internals inside a valve body. In the absence of this precaution, the generated heat during the welding of the valve's body to a pup or transition piece may damage soft materials such as the seals in the valve. Prior to shipping the valve to the construction yard and welding it to the pipeline, the valve is welded to the pup or transition piece. The forged pipe, referred to as the "pup piece," is used to divert welding heat away from the internal components of the valve, especially the sealing components. As an example of a forged fitting, a transition piece is welded to the valve body from one side and the pup piece from the other side. As shown in Figure 7.10, a 38″ Class 1500 top entry ball valve body for the first pipeline valve on the oil export line is welded to pup and transition pieces. An image shows the welding of three components: the valve body in A352 LCC (carbon steel) on the right, the transition piece in ASTM A694 F52 (high-strength low-alloy steel) in the middle, and the pup piece in ASTM A182 F51 (22Cr duplex) on the left. Transition pieces should be welded between the pup piece and the valve body if the thickness differential between the valve and pipe (pup piece) exceeds 50% (refer to ASME B31.8 code). The pipeline and the pup piece have the same thickness. Specifically, the valve and the pipeline connected to it are each 103-mm thick and 60-mm thick, respectively. This transition piece consists of a 38″ forged pipe of 103 mm thickness with one end facing the valve and 60 mm thickness with the other facing the pup piece. With smaller pipeline valves (e.g., 20″), the thickness difference between the valve and pipe is less than 50% of the valve and pipe thickness values, so such welding can be performed without a transition piece. All three valve bodies in this case study are made of LTCS castings, ASTM A 352 LCC. The valves and pipeline are connected by six weld joints. Other than the downstream side of the third valve connected to the subsea pipeline made of API 5LX65 material, welds are performed to a pipeline made of 22Cr duplex according to the picture and schematic given in Figures 7.10 and 7.11.

It was not proposed to use a 22Cr duplex transition piece since welding a 22Cr duplex transition piece to a 103-mm carbon steel valve would produce high heat input and possibly create a brittle phase in the 22Cr duplex transition piece. In fact, duplex stainless steel can make structural transformation, leading to the formation of the sigma phase in response to high temperature. In addition, Sigma intermetallic compounds can reduce the mechanical strength and corrosion resistance of duplex stainless steels. It would be less risky and easier to connect an ASTM A694 F52 transition piece with a 22Cr duplex pup piece with a thickness of 60 mm. However, welding a high-strength low-alloy steel pup piece made in an API 5LX65 to a 22Cr duplex pup piece remains challenging due to the low fusion risk associated with the hard materials. Consequently, F52 is chosen as the transition piece since this material is not very hard, but has sufficient mechanical strength. There are challenges with welding F52 transition pieces and 22Cr duplex pup pieces in 60 mm thickness where F52 materials require post weld heat treatment (PWHT) to release residual stresses

FIGURE 7.10 The first ball valve body in 38″ CL1500 on the oil export pipeline after the pup and transition pieces have been welded together. (Photo by the author.)

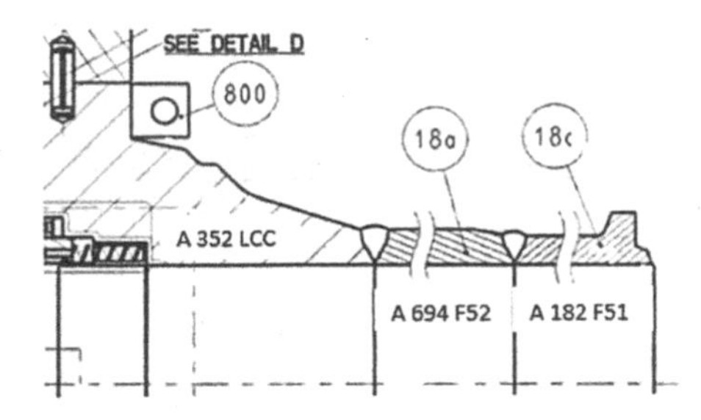

FIGURE 7.11 Schematic of the valve body, transition piece, and pup piece connections as per the previous illustration.

in accordance with ASME B31.3. In contrast, PWHT is not recommended for 22Cr duplex due to the risk of sigma phase formation often associated with 22Cr duplex. The solution is to butter the F52 pup piece material with Inconel 625 and apply PWHT prior to welding it to the 22Cr duplex pup piece. It is then possible to use Inconel 625 filler (ERNiCrMo3) to weld the buttered F52 transition piece to the F51 duplex pup piece as shown in Figure 7.12 without applying PWHT. Material engineers may propose using alloy 59 without niobium instead of Inconel 625 to achieve better welding quality between the pup and transition pieces. Figure 7.12 illustrates how both sides of the first and second pipeline valves as well as the upstream side of the third pipeline valve are welded to 22Cr duplex pipeline. According to the schematic shown in Figure 7.13, the right side (downstream) of the third pipeline valve is welded to the subsea pipeline. Similar to the first and second pipeline valves, the body of the third pipeline valve is made of A352 LCC material. To weld the valve body to the pup piece, a low-alloy steel transition piece made of A694 F52 is used as usual. During the welding process, the pup piece was selected in low-alloy steel (A694 F65) instead of 22Cr duplex (A182 F51), because the pup piece should be

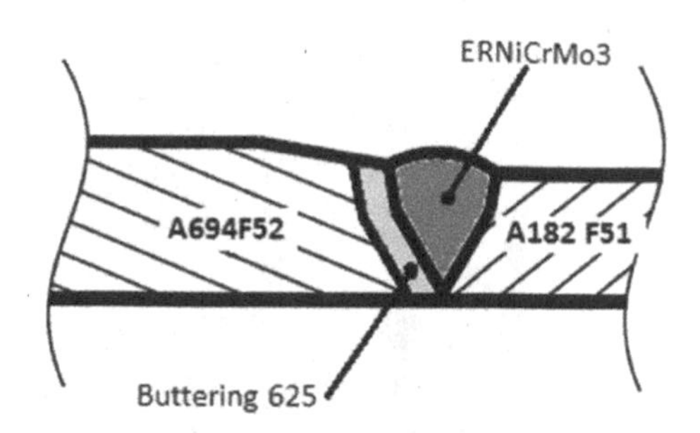

FIGURE 7.12 Welding schematic between F52 transition piece and F51 pup piece through Inconel 625 buttering. (Photo by the author.)

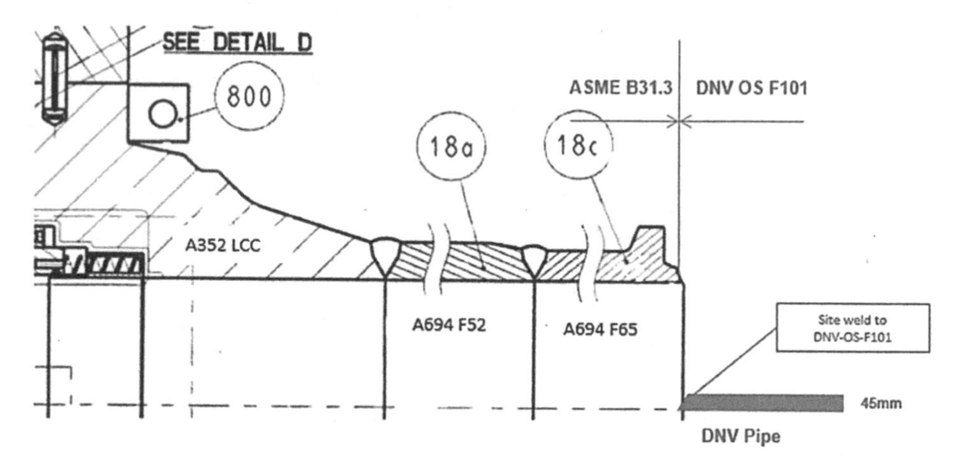

FIGURE 7.13 A schematic for welding the right side (downstream) of the third pipeline valve to the subsea pipeline in accordance with DNV standards. (Photo by the author.)

made from the same material as the pipeline. From the right side, the third valve is welded to a subsea pipeline made of API 5LX65.

In general, the lengths of the pups and the transition pieces are determined by the size of the valve and the pipeline. As a general rule, they should be long enough to protect the valves' internal components from the heat generated during the onsite welding process. Table 7.1, the middle column, presents the length values of a pup piece if the valve is connected to the pipeline without a transition piece. If both the pup and transition piece are used, the values in the table should be applied to the total length of the pup and transition piece taken together. This is because the length of the pup and transition piece should be the same.

In Figure 7.14, we depict a 38″ CL1500 ball valve, which can be installed as the first or second pipeline valve after the pig launcher, as shown in Figure 7.9. The pup and transition pieces each measure 400 mm in length, which results in a total length of 800 mm on either side of the valve. It should be noted that the pup and transition pieces' lengths do not always follow the values in the table. For example, Figure 7.15 illustrates a 38″ CL1500 ball valve, which is the third pipeline valve after the pig launcher.

TABLE 7.1
Length Determination for a Pup and Transition Piece

	Only Pup Piece	Pup Plus Transition Pieces
Valve Size	**Pup Piece Length**	**Pup Plus Transition Piece Length**
Maximum 8″ NPS or DN200	8″ or 200mm	8″ or 200mm (pup piece length = transition piece length = 100 mm)
10″ to 20″ NPS (DN250 to DN500)	NPS (DN)	NPS (DN) pup piece length = transition piece length = $\frac{NPS(DN)}{2}$
22″ NPS (DN550) and larger	32″ or 800mm	32″ or 800mm (pup piece length = transition piece length = 400 mm)

FIGURE 7.14 A 38″ CL1500 pipeline ball valve with 800 mm extensions on each side. (Photo by the author.)

In accordance with the table, the length of the pup and transition pieces on the left side of the valve is 800 mm. Nevertheless, the pup piece on the right side connected to the subsea pipeline is extended from 400 mm to 1780 mm so that it can weld directly to the subsea pipeline. Therefore, the total length of the pup and transition piece on the right side of the valve is 400 mm plus 1780 mm, which is 2180 mm. Assume that the pup piece does not extend any longer. In that case, an API 5LX65 pipe with a length of 1380 mm should be welded to the valve pup piece from one side, and the subsea pipeline from the other side, thus increasing the number of welding and onsite activities.

FIGURE 7.15 A 38″ CL1500 pipeline ball valve with extended pup piece on the right. (Photo by the author.)

7.3.3 Weld Joint NDTs

Non-destructive testing is a process of inspecting, testing, and evaluating materials, components, or discontinuities without destroying them. This means that the part that has undergone the NDT can be used after the test. Nowadays, in the modern industry, NDT is used during manufacturing, fabrication, and inspection to ensure product integrity and reliability and control the manufacturing and fabrication process effectively, lower the cost due to failure during operation, and improve quality. The other essential consideration is related to the inspection and NDTs after conducting circumferential welds between the valves' bodies, pup, and transition pieces. Five types of NDTs or inspections are typically performed on 100% of the circumferential welds between the valves' bodies, pup, and transition pieces: visual testing (VT), magnetic particle test (MT), liquid penetration test (PT), radiography (RT), and ultrasonic examination (UT). A visual test is the easiest and cheapest means of inspection to detect welding defects on the surface of a material. It can be done with a naked eye or with a magnifier. This method of testing alone is not accurate. Therefore, it is typically combined with other methods, such as MT, PT, RT, or UT. One or more magnetic fields are used in magnetic particle testing to find welding defects on or near the surface of ferromagnetic materials. Figure 7.16 illustrates a liquid penetration test between the pup and transition pieces belonging to a 38″ CL1500 pipeline ball valve installed on an oil export line. Three types of liquid are used in the PT test to detect only surface defects; the first is a red liquid called penetrant (see Figure 7.16) that is soaked into flaws. The second called developer is a white liquid that pulls the penetrant back out of a defect to allow the inspector to see the defect, and the last one is a cleaner.

Contrary to VT, MT, and PT, which detect defects on the surface, RT and UT are known as volumetric tests, meaning that they can detect defects and discontinuities

FIGURE 7.16 A liquid penetration test between a pup and transition piece of a 38″ CL1500 pipeline ball valve. (Photo by the author.)

below the surface. An industrial radiography test involves exposing the weldment or any other test object to radiation, which passes through the object being inspected. X-rays are widely used for RT examination; however, they cannot be used for thick components, so a gamma ray may be used instead in these cases. During the test, the material is placed between the source of radiation and the film. The results of the RT are displayed on the film, which can highlight defects such as porosity in the welds. Another type of volumetric NDT is the ultrasonic test, in which a high frequency ultrasonic wave or sound is propagated inside a material. In this test, an ultrasonic beam is injected into the material being tested. Except when it is interrupted or deflected by a defect or discontinuity, the beam travels inside the object without loss.

7.3.4 Pipeline Valves Welding

There are four types of welding performed by welders on pipeline valves:

1. In this chapter, we have discussed the welding between the valves' bodies, pups, and transition pieces.
2. Tungsten carbide, as its name suggests, consists of tungsten and carbide atoms that have very high strengths and hardness. In order to reduce galling and wear during valve operation, tungsten carbide is applied to the contact surfaces between the closure member and the seats. Galling occurs when two surfaces slide over each other, causing them to wear. Tungsten carbide can be applied by two methods: high velocity oxygen flux (HVOF) thermal spray coating in which oxygen and fuel are mixed to produce the required heat to assure a low porosity and high bond strength between the tungsten carbide and a substrate. Alternatively, the tungsten carbide can be applied by welding.

3. Weld cladding usually refers to the application of a relatively thick layer ($\geq$3 mm or $\frac{1}{8}$") of a CRA like Inconel 625 to provide a corrosion-resistant surface. There are seat pockets and sealing areas machined as grooves on the pipeline valves, exposed to crevice corrosion. Crevice corrosion is a type of corrosion that occurs in confined spaces where fluid flow is limited. The seat pocket and sealing areas are crevices, where fluid can be trapped and cause crevice corrosion. In short, crevice corrosion can be defined as a breakdown on a metal or alloy with a tight crevice due to a corrosive solution or fluid. In this case, all machined grooves in the valve body and bonnet, including sealing areas and seat pockets, shall be weld overlaid or cladded with at least 3 mm Inconel 625 to prevent crevice corrosion.
4. Casting is one of the essential manufacturing milestones that is discussed in detail in Chapter 6. A number of tests and quality control activities are carried out on the produced casting. NDTs, such as visual inspection, magnetic particle inspection, and ultrasonic examination, are applied following heat treatment to detect possible defects and discontinuities. Defects found during NDT and not accepted are repaired by welding, or ***"weld repair."***

7.3.5 Welding Qualification at the Construction Yard

In this case study, two types of welding are performed in the construction yard between valves and the pipeline; one is between A694 F52 pup pieces and a 22Cr duplex pipeline, and the second is between an A694 F65 pup piece on the right side (downstream) of the third pipeline valve and an API X65 subsea pipeline. Before welding in the construction yard, welders need to qualify both types of welds between the valve pup pieces and the pipeline. As a result, for welding qualification activities in the construction yard, a valve purchaser should order at least two additional pup piece rings with the same materials and heat numbers as pipeline valve pup pieces. The welding qualification is performed by a qualified welder in accordance with the welding procedure specification (WPS), and the weld parameters are tested and recorded according to the procedure qualification record (PQR). A PQR is used as evidence to demonstrate that a WPS can be used to produce acceptable welding, and that production welding can begin following the evaluation.

QUESTIONS & ANSWERS

1. What is the preferred end connection for pipeline valves?

   A. Flanged
   B. Hub and clamp
   C. Butt weld
   D. Thread

   Answer) Pipeline valves have butt weld end connection, so option C is the correct answer.

2. What is the correct sentence about butt weld?

   A. The main standard for butt weld connection is ASME B16.24.
   B. The average value of the root face in the butt weld connection is 1.6 mm.
   C. The bevel end and butt weld configurations for a pipe with a wall thickness of 19 mm and 25 mm are the same.
   D. Narrow gap welding requires more welding consumables compared to the standard butt weld connection.

   Answer) The leading standard for butt weld connection is ASME B16.25, so option A is incorrect. Option B is the correct answer. Option C is incorrect since bevel ends and butt welds are configured differently for pipe joints with less than 22 mm and more than 22 mm thickness values. Option D is not correct either because narrow gap welding requires fewer welding consumables than a standard butt weld connection.

3. Select the correct sentence.

   A. Pipeline valves with a thickness of 75 mm should be welded to pipes with a thickness of 65 mm. A pup and transition piece are not required for such welding.
   B. Pipeline valves are always made from the same material as the pipe they are connected to.
   C. A transition piece has an equal thickness on both sides.
   D. Even if the materials of a pipeline valve and the connected pipeline are not identical, the material of the pup piece is equal to the pipeline.

   Answer) Pipeline valves and the connected pipelines have a thickness difference of 10 mm, which is less than 50% of the valve and pipeline thicknesses. Therefore, a transition piece is not needed for such welding. As a minimum, pipeline valves must be permanently welded to the pipelines through a couple of pup pieces at both ends. Therefore, option A is incorrect. Option B is also not appropriate since the material of the pipeline valve may be different from the material of the pipeline it is attached to. Option C is incorrect since the thickness of a transition piece differs on both sides. The correct answer is D.

4. A 20" CL1500 gas export pipeline with a forged carbon–steel body (ASTM A105) is connected to a 45-mm thick 22Cr duplex pipeline on one side and to an API 5LX65, 50-mm thick submarine pipeline on the other. Which of the following sentences is correct?

   A. Welding between the valve and pipeline requires the use of both pup and transition pieces.
   B. The pup pieces used to weld the valve to the pipelines are made of 22Cr duplex.
   C. The pup pieces have a thickness of 45 mm on one side and 50 mm on the other side.
   D. All three sentences are wrong.

   Answer) Option A is incorrect because no transition piece is required to weld the gas export pipeline valve to the pipeline. Moreover, the thickness differences between the valve and connected pipeline are 15 mm and 10 mm, respectively. These thickness differences are less than half the thickness of both pipeline and valve. Therefore, the valve can be welded to the pipeline by a couple of pup pieces on both sides. Option B is also incorrect, as the material of pup pieces should be the same as that of the connecting pipeline. Thus, one pup piece is forged from 22Cr duplex, and the other is forged from ASTM A694 F65, a forged alloy that is compatible with API 5LX65. Option C is the correct response. Option D is incorrect since option C is the correct response.

5. Find the correct sentence about welding, material selection, and non-destructive tests (NDTs).

   A. The joint efficiency is 90% if the weld and base metal strength are 18,000 and 20,000 psi, respectively, and the weld is radiography tested 100%.
   B. PWHT is not recommended for duplex materials due to the risk of sigma phase formation.
   C. Liquid penetration is a type of volumetric NDT.
   D. Pipeline valves can be made of 22Cr duplex.

   Answer) As a result of applying the radiography test, option A is invalid, and the joint efficiency is 100%. The correct response is option B. Option C is incorrect, as liquid penetration is only used to detect surface defects. However, volumetric NDTs such as radiography (RT) and ultrasonic examination (UT) can detect defects that lie beneath the surface. Option D is incorrect. Pipeline valves are manufactured from carbon or low-temperature carbon steel materials. Therefore, option B is the correct choice.

6. What type of welding is applied to pipeline valves to prevent crevice corrosion?

   A. Welds between the pup pieces and bodies of the valves
   B. Tungsten carbide overlay on closure members and seats
   C. Inconel 625 cladding on the machined grooves inside valves
   D. Weld repairs on the casting defects

   Answer) Option C is the correct answer. A weld between the pup pieces and the valve bodies is performed to reduce the distance between the welds conducted in the construction yard and the valve internals, especially the seals, and to protect these against the heat produced during the welding process. Valve internals are coated with tungsten carbide to prevent wear and galling. Welded repairs are performed to correct casting defects.

BIBLIOGRAPHY

1. American Petroleum Institute 609. (2004). *Butterfly valves: Double flanged lug and wafer*, (6th ed.). Washington, DC: ASME.
2. American Society of Mechanical Engineers (ASME) B1.20.1. (2013). Pipe Threads, General Purpose, Inch. New York, NY: ASME.
3. American Society of Mechanical Engineers (ASME) B16.11. (2016). *Forged fittings, socket-welding and threaded*. New York, NY: ASME.

4. American Society of Mechanical Engineers (ASME) B16.5. (2017). Pipe flanges and flanged fittings: NPS ½″ through NPS 24 Metric/inch standard. New York, NY: ASME.
5. American Society of Mechanical Engineers (ASME). (2017). Butt welding ends. ASME B16.25. New York, NY: ASME.
6. American Society of Mechanical Engineers (ASME) B16.34. (2017). *Valves–Flanged, threaded, and welding end.* New York, NY: ASME.
7. American Society of Mechanical Engineers (ASME) B16.47. (2017). *Large diameter steel flanges: NPS 26″ through NPS 60″ metric/inch standard.* New York, NY: ASME.
8. American Society of Mechanical Engineers (ASME) B31.3. (2012). *Process piping.* New York, NY: ASME.
9. American Society of Mechanical Engineers (ASME) B31.4. (2019). *Pipeline transportation systems for liquids and slurries.* New York, NY: ASME.
10. American Society of Mechanical Engineers (ASME) B31.8. (2018). *Gas transmission and distribution piping systems.* New York, NY: ASME.
11. American Society of Mechanical Engineers (ASME). (2019). Rules for construction of pressure vessels. Boiler and Pressure Vessel Code. ASME Section VIII Div.01. New York, NY: ASME.
12. Assured Automation. (2019). Types of valve end connections. [Online]. Retrieved from https://assuredautomation.com/actuated-valve-training/types-of-valve-end-connections.php [access date: 28th December 2021].
13. Cristini B., Sacchi B., Guerrini E., and Trasatti S. (2010). Detection of sigma phase in 22Cr duplex stainless steel by electrochemical methods. *Russian Journal of Electrochemistry*, 46(10), 1094–1100.
14. Equinor. (2014). Piping engineering. Technical Requirement (TR 2014). Revision 4.
15. Engineering Edge. (2019). Welded joint efficiency table recommendations. [Online]. Available from: https://www.engineersedge.com/weld/welded_joint_efficiency_14419.htm [access date: 31st December 2021].
16. International Organization for Organization (ISO). (2012). Petroleum and natural gas industries—Compact flanged with IX seal ring. ISO 27509. 1st ed. Geneva, Switzerland.
17. Nesbitt B. (2007). *Handbook of valves and actuators: Valves manual international*, (1st edition). Oxford, UK: Elsevier.
18. NORSOK. (1999). Piping and Valves. NORSOK L-001, Revision 3. Lysaker, Norway.
19. NORSOK. (2003). Compact flanged connections. NORSOK L-005, Revision 1. Lysaker, Norway.
20. Smit P. & Zappe R. W. (2004). *Valve selection handbook*, (5th ed.). New York, NY. USA: Elsevier.
21. Sotoodeh K. (2015). Top entry export line valves design considerations. *Valve World Magazine*, 20(05), 55–61.
22. Sotoodeh K. (2016). Piping and valve materials for offshore use: Part 1. *Stainless Steel World Magazine*, 28, 40–44.
23. Sotoodeh K. (2017). Duplex & super duplex piping & valves in offshore applications. *Stainless Steel World Magazine*, 29, 46–50.
24. Sotoodeh K. (2018). Why are butterfly valves a good alternative to ball valves for utility services in the offshore industry? *American Journal of Industrial Engineering*, 5(1), 36–40. https://doi.org/10.12691/ajie-5-1-6
25. Sotoodeh K. (2020). Manifold technology in the offshore industry. *American Journal of Marine Science*, 8(1), 14–19.
26. Sotoodeh K. (2021). Dissimilar welding between piping and valves in the offshore oil and gas industry. *Welding International, Taylor and Francis*. https://doi.org/10.1080/09507116.2021.1919495

27. Sotoodeh K. (2021). *A practical guide to piping and valves for the oil and gas industry*, (1st edition.) Austin. USA: Elsevier (Gulf Professional Publishing).
28. Sotoodeh K. (2021). Subsea valves and actuators for the oil and gas industry. Elsevier (Gulf Professional Publishing). 1st edition. Austin. USA.
29. Sotoodeh K. (2022). Piping engineering: preventing fugitive emission in the oil and gas industry. Wiley.1st edition. USA.

8 Actuation

8.1 INTRODUCTION

Except for check and pressure relief valves, every valve requires a mechanism for operation. Depending on their operation method, valves are classified into two categories: manual and actuated. Manually operated valves or manual valves are those that are operated by levers, handwheels, and gear gearboxes. Figure 8.1 illustrates a manually operated valve with a handwheel and a gearbox that is operated by a human operator.

A valve actuator is a mechanical or electrical component mounted on top of the valve (Figure 8.2) that is used to move and control the valve's opening and closing. The actuators function by converting external energy sources, such as air, hydraulic power, or electricity, into mechanical motions. Therefore, actuators are considered one of the most vital components of valve automation. Actuators make it possible for valves to be automated in such a way that no human interaction is necessary in order to operate them. Actuators are more popular today than they were in the past for many reasons, including the reduced need for personnel to operate the valves, precision control of valve operation, simplicity, safety, speed, and reliability of valve operation. Valves that are located in remote locations or require frequent operation require actuators. According to the data published by European Industrial Forecasting

FIGURE 8.1 An operator operates a manual valve in a plant. (Courtesy: Shutterstock.)

DOI: 10.1201/9781003343318-8

FIGURE 8.2 A pipeline valve with a hydraulic actuator at the top. (Photo by the author.)

Ltd. (EIF) in 2015, 75% of the valves in the oil and gas industry are automated by actuators. Actuation of valves in the oil and gas industry is much higher than the actuation rate across all industries, which is reported to be 30%. Valve quality is primarily determined by selection of proper materials and metallurgy of the valve components, their mechanical strength, especially when pressure-containing parts are involved, appropriate machining, and tight tolerances, for example. However, the valve's performance, such as its ability to open and close, also known as cycling, is largely dictated by the actuators. The actuator can be powered by air, hydraulic oil, electricity, or by a combination of hydraulic power and electricity, which is explained in the next section.

8.2 ACTUATORS' SOURCES OF POWER

There are three types of actuators: pneumatic, hydraulic, and electrical, which are discussed in this section.

8.2.1 Pneumatic Air

These pneumatic actuators are powered by compressed and pressurized air with a pressure value of around 5 to 9 bar, which is the primary power source for these actuators. Both linear and quarter-turn (rotary) valves may be controlled by pneumatic actuators. The advantage of pneumatic actuators is that they operate with air, which is a safe and environmentally friendly alternative to hydraulic oil. It is not harmful to the environment when compressed air leaks from the actuator; however, it does have other negative consequences, such as increasing the cost of operation. The main limitation of pneumatic actuators is their inability to supply sufficient torque or force to operate large-size and high-pressure class valves, such as pipeline valves. Therefore, pneumatic actuators should be replaced by hydraulic actuators for pipeline valve because of large sizes and high-pressure classes.

8.2.2 Hydraulic

The hydraulic actuator converts hydraulic oil pressure into mechanical movement. The working principle of a hydraulic actuator is similar to that of a pneumatic actuator, with the difference being the source of fluid power. To operate hydraulic actuators, pressurized hydraulic oil, with a pressure of 160 to 200 bar, is used instead of pneumatic air. Consequently, hydraulic actuators can produce a greater force or torque for valve operation compared to pneumatic actuators. In fact, hydraulic actuators are a better choice for valves that call for a high amount of torque or force to operate, such as pipeline valves. A larger valve in a high-pressure class typically requires a greater amount of torque. The advantages of hydraulic actuators include:

- As a result of the higher oil pressure than that of air, hydraulic actuators are more compact than pneumatic actuators. Since the oil pressure in a hydraulic actuator is higher than the air pressure in a pneumatic actuator, less oil is required to create the same force as air.
- Hydraulic actuators are more precise than pneumatic actuators since the oil is not compressible.
- Hydraulic actuators provide very high-speed operation for large-size and high-pressure class valves. The speed of hydraulic actuators is higher than that of both pneumatic and electrical actuators.

The disadvantages of a hydraulic actuator are summarized as follows:

- The high pressure of hydraulic fluid handling requires a high level of safety and precaution.
- Unlike with air, the leakage of hydraulic oil negatively impacts the environment.
- The control system of a hydraulic actuator is typically a large box.

8.2.3 Electrical

In electrical actuators, electric power is used to operate the valves. Electrical motors may use AC or DC voltage. Electrical actuators are also known as motors, which are often coupled with a gearbox if a higher torque or force value is required to operate a valve. Consider, for example, selecting an electrical actuator for a pipeline valve with a 30" diameter and a pressure class of 1500 equal to 250 bars. If this is the case, the motor must be upgraded with a gearbox to provide higher torque for the valves. Among the main advantages of electrical actuators are that they are inexpensive, easy to handle, compact, and environmentally friendly. Electrical actuators, with the exception of a few configurations equipped with spring or hydraulic power, cannot provide a fail-safe position. Alternatively, the electrical actuators can perform a fail-as-is or a fail-at-last position. The disadvantage of this type of actuator is that it is unable to provide the force required for rapid valve operation, making it unsuitable for emergency shutdown (ESD) valves.

8.3 ACTUATOR CHOICES FOR PIPELINE VALVES

Three choices of actuators are typical for these valves: scotch and yoke, linear, and electrical that are explained in detail in the following section.

8.3.1 Scotch and Yoke Actuator

Scotch and yoke actuators are suitable for quarter-turn valves such as ball valves. Actuators of this type can be operated by hydraulic or air power and can be either single-acting (spring-return) or double-acting. Single-acting actuators are pressurized by air or hydraulic oil from one side and return to the fail-safe mode by spring force from the other. Figure 8.3 illustrates a spring-return scotch and yoke actuator that converts linear or reciprocating motion into a rotary or quarter-turn motion. Through a slot or pin, the shaft in the middle of the actuator is directly connected to the yoke. The air or hydraulic fluid enters the actuator from the left side and drives the shaft forward. Through the slot or pin, the shaft's linear motion is transferred to the yoke. By rotating the yoke and shaft 90 degrees, the valve can be moved between open and closed positions. Figure 8.3 depicts a spring-return actuator and a connected valve that can either be fail-safe open or fail-safe closed. In a fail-safe open design, the pressure of air or hydraulic oil applied to the left side of the actuator overcomes the spring force on the right side of the actuator and makes the actuator and the connected valve closed. If, on the other hand, the air or hydraulic oil supply to the actuator stops, the spring force will be greater than the air or hydraulic oil force, which will result in the valve being opened. In a fail-safe closed design, air or hydraulic pressure opens the actuator and connected valve, while spring tension closes them. In contrast to single-acting actuators, double-acting actuators have a springless operation and are actuated by both sides of the fluid. As a result of their springless design, double-acting actuators cannot provide a fail-safe mode in case of failure of the supply fluid or air. The failure mode of double-acting actuators is

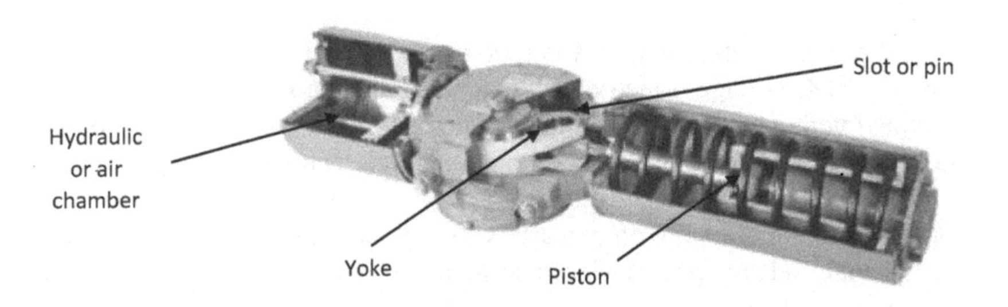

FIGURE 8.3 Scotch and yoke actuator. (Photo by the author.)

stay-in-position or fail-as-is. Generally, double-acting actuators are more compact than single-acting actuators and can produce greater forces.

8.3.2 Linear Actuator

Linear actuators like scotch and yoke can work with either air or hydraulic systems. However, unlike scotch and yoke actuators, they can only generate force in a linear or straight-line suitable for operating through conduit gate (TCG) valves. A linear actuator illustrated in Figure 8.4 is usually installed vertically on the top of a TCG valve. The hydraulic oil enters the chamber from the top and overcomes the spring force, so the piston rod moves down, and consequently, the movement of the stem rod is transferred to the valve. Linear piston actuators could be either "single-acting," which is also called "spring-return," or "double-acting." with exact mechanisms explained before for scotch and yoke actuators. Single-acting actuators, as explained before, are pressurized by the hydraulic oil or air from one side and return to the fail-safe mode from another side by the spring force. However, double-acting actuators have a springless operation and are pressurized by both sides' hydraulic oil or air. Due to the springless design, double-acting actuators cannot provide a fail-safe mode in case the air or hydraulic oil supply fails. The failure mode of double-acting actuators is fail-as-is or staying-in-position.

8.3.3 Electrical Actuator

Electrical actuators can provide rotary motion for quarter-turn valves such as ball valves and linear motion for linear valves like TCG valves because the ball and TCG

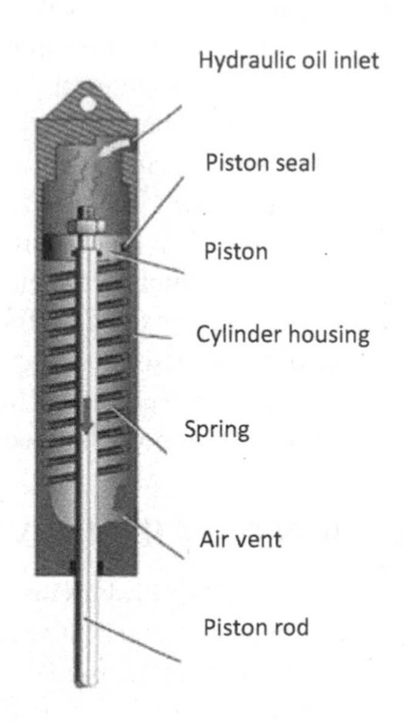

FIGURE 8.4 Linear spring-return hydraulic actuator. (Photo by the author.)

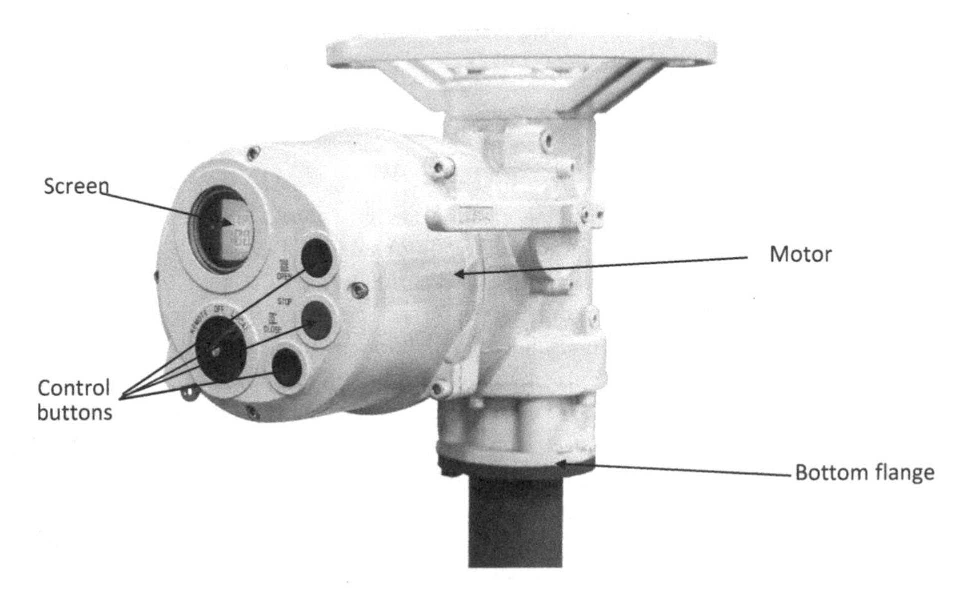

FIGURE 8.5 An electrical actuator and its parts. (Courtesy: Shutterstock.)

valves are a couple of choices for pipeline valves. Figure 8.5 illustrates an electrical actuator, including motor, screen, control buttons, and bottom flange. The motor is a part that is powered by electricity and converts the electricity to mechanical force. Electrical actuators typically have a screen on site that shows the valve opening percentage. For example, Figure 8.5 illustrates the 100% opening of the valve on the screen of the electrical actuator. There are three buttons in the picture: one for opening, one for closing, and one for stopping the actuator. These three buttons work when an operator is operating the actuator locally. Local operation means that the operator is standing next to the actuator and is not typically permitted without a signal from the control room. A control room is a large operation room with facilities used to monitor and control the plant. A big black button on the electrical actuator illustrated in the figure can be used to switch between local and remote modes of operation. Remote operation of the actuator can be performed from the control room. There is a handwheel on an electrical actuator for manual operation, which allows the operation of the actuator manually by the operator when the actuator is off. The bottom flange is the interface or connection point between the actuator and the valve.

8.4 ACTUATOR SELECTION FOR PIPELINE VALVES

Large and high-pressure ball valves and TCG valves are installed on oil or gas export pipelines after pig launchers. In Chapters 1 and 3, it was explained that three valves are typically installed on each pipeline after the pig launcher. The last or third valve in a pig launcher is the only one equipped with an ESD and fail-safe closed function. In the event that something is wrong on the subsea pipeline, the third valve, which is normally opened, is closed rapidly and automatically. The proper choice

TABLE 8.1
An Overview of the Selection of Actuators for Pipeline Valves

Pipeline valve 1: Ball valve Actuator type: Electrical actuator	Pipeline valve 2: Ball valve Actuator type: Electrical actuator	Pipeline valve 3: Ball valve Actuator type: Hydraulic spring-return scotch and yoke actuator
Pipeline valve 1: TCG valve Actuator type: Electrical actuator	Pipeline valve 2: TCG Actuator type: Electrical actuator	Pipeline valve 3: TCG Actuator type: Hydraulic spring-return linear actuator

of actuator for the third valve is a spring-return or single-acting hydraulic actuator. Pneumatic actuators, as mentioned earlier in this chapter, are not suitable since they are unable to deliver the necessary force for the pipeline valve to operate. Electrical actuators cannot provide the fast operation and fail-safe closed function required for valves with ESD functions. The first valve after the pig launcher is a normally closed (NC) valve that should be opened when the pig is shot into the pipeline and closed afterward. There is no ESD mode associated with the opening and closing of the first valve after the pig launcher, and its actuation is merely for ease of operation. If the first valve is operated manually through the use of a handwheel and gearbox, a human operator may need to spend a considerable amount of time (e.g., 20 minutes) opening the valve from the closed position. However, adding an electrical actuator or motor can reduce this time to only a few minutes. Therefore, an electrical actuator is required for the first valve after the pig launcher to facilitate its operation. It is evident that an electrical actuator is the most suitable choice if ease of operation is required. The second valve after the pig launcher serves as a backup for the first valve, so both have identical electrical actuators. Contrary to the first valve, the second valve after the pig launcher always remains open (normally open [NO]). For the last valve after the pig launcher, a linear actuator is required if it is a TCG valve. A scotch and yoke actuator is required if it is a ball valve. Electrical actuators are selected for the first and second valves after the pig launcher, regardless of whether they are ball or TCG valves. Table 8.1 summarizes the selection of actuators for pipeline valves. The three valves after the pig launcher are all selected from one type of valve in a project (e.g., three ball valves or three TCG valves).

8.5 ACTUATOR SELECTION CASE STUDY

Here is a case study of hydraulic actuator selection for the third pipeline valve after the pig launcher in a recent offshore project. As part of the Johan Sverdrup project, this case study focuses on the challenges of choosing an actuator for a 38" pipeline valve. One of the five largest oil fields on the Norwegian continental shelf is the Johan Sverdrup project that is known as one of Norway's most important industrial projects for the next 50 years. It is expected to produce 2.7 billion barrels of oil from the field. Between 2015 and 2025, the project development could generate at least 150,000 man-years of employment in Norway. The field is located 160 kilometers west of Stavanger, a city in the western part of Norway. There are four platforms in the project: riser, drilling, process, and living quarters (see Figure 8.6).

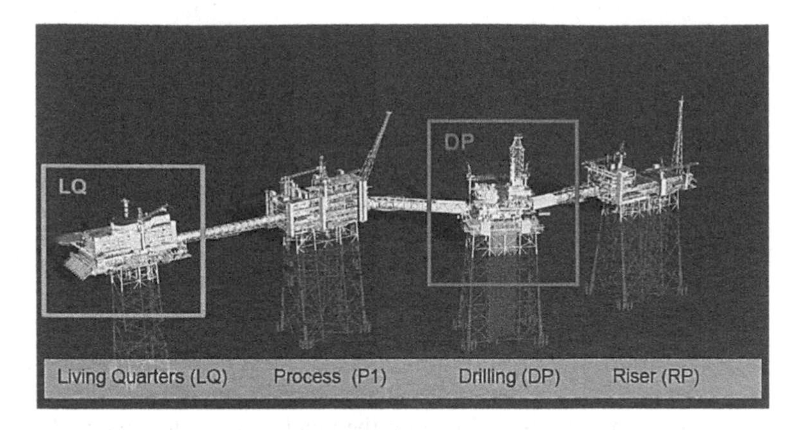

FIGURE 8.6 Platforms for Johan Sverdrup project.

FIGURE 8.7 38" Class 1500 ESD pipeline ball valve with double-acting hydraulic actuator. (Photo by the author.)

There is an oil export pipeline of 38" size and a CL1500 to transport oil from the riser platform to the shoreline. Three valves are located on the oil export line on the platform, upstream (before) of the connection between the topside and subsea pipeline and downstream (after) of the pig launcher. There is an emergency shutoff valve at the end of the oil export line (see Figure 8.7), which is connected to the subsea line. As a matter of fact, the valve should be closed in the event of a pipeline failure in order to prevent loss of production and maintain health, safety, and environment (HSE) concerns. The valve is approximately 67 tons in weight and should be equipped with a hydraulic actuator because it is very large in size and high-pressure class, and it should also have an ESD function.

The first issue was that the spring-return actuator for this valve could weigh nearly 15 tons. The addition of approximately 15 tons of weight from the actuator to the huge valve that weighs approximately 82 tons would make the valve complex to handle, both in terms of actuator installation on the valve and transportation of the valve and actuator assembly. In addition, it should be noted that the valve and actuator were assembled at the valve manufacturer's factory, located in South France, and then shipped to South Korea. It was determined that the solution was to change the actuator selection from a single-acting to a double-acting hydraulic type, resulting in a weight reduction of approximately 10 kilograms. Another problem associated with spring-return hydraulic actuator selection was the failure to maintain the safety factor for actuator sizing. The end user (client) requested that a safety factor of at least 2 be applied to actuator sizing in both closed and open positions. The safety factor is defined as the minimum ratio of the actuator torque or force to the corresponding valve torque or force. The safety factor is calculated by using Equation 8.1. A safety factor of a minimum of 2 means that the actuator should produce torque values at least twice of those required to operate the valve. If such a large and high-pressure valve is to be closed, a very strong and high torque spring should be selected for the actuator in order to ensure a safety factor of at least 2. In addition, if a solid and high torque spring is selected, a high hydraulic oil action force will be required to operate the actuator. As a result, the actuator manufacturer may face challenges in maintaining a safety factor of two for both spring and oil forces. Despite the large and solid spring force in the actuator, the oil force could overcome it and open the valve with a safety factor of only 1.7 to 1.8.

Equation 8.1 Actuator safety factor calculation

$$\text{Safety factor of actuator} = \frac{\text{Torque or force produced by an actuator}}{\text{Torque or force required for valve operartion}}$$

Five torques are related to each valve and connected actuator, including the pipeline valves:

1. ***Break to open (BTO):*** This torque, also called breakaway torque, is measured when the valve is closed, and the closure member (e.g., ball in a ball valve) begins to open against just one seat under pressure.
2. ***Running torque (RT):*** The torque of the valve when the closure member opens at approximately 35° to 45° for a ball valve or 40% to 50% of the gate full opening for a gate valve.
3. ***End to open (ETO):*** The torque required to fully open the valve when the closure member opens at the 80° or 90% closed position.
4. ***Break to close (BTC):*** When the valve is in the fully open position, the torque required to break the open position of the valve to close the valve.
5. ***End to close (ETC):*** The torque required to fully close the valve when the valve is about to close.

For safety factor calculations, it should be noted that each valve torque condition (i.e., BTO) should be compared with the relevant actuator torque value (BTO). As mentioned earlier, the safety factor for actuator sizing is calculated in five different positions. Table 8.2 contains torque values provided by a valve manufacturer for a 38" ball valve CL1500, as well as actuator torque values based on a safety factor of 2 under all operating conditions. Because actuators cannot be designed in such a way that they have a safety factor of precisely 2 for all torque conditions, the actuator torques listed in the table are theoretical values to maintain a safety factor of 2. In relation to the industrial experiences of the author, it is essential to note that safety factors for electrical actuators are typically lower and may even be as low as 1.5.

TABLE 8.2
Calculation of the Actuator Safety Factor

	Valve torque values (Nm)			
BTO	**RUN**	**ETO**	**BTC**	**ETC**
348939	61510	139794	348939	139794
	Actuator torque values (Nm)			
BTO	RUN	ETO	BTC	ETC
697878	123020	279588	697878	279588
	Safety factor values			
2	2	2	2	2

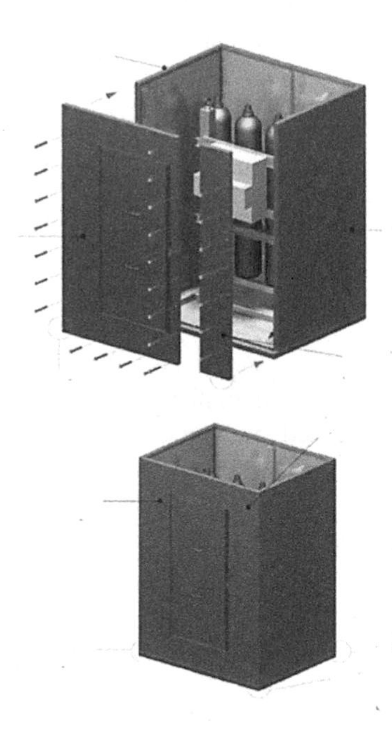

FIGURE 8.8 An accumulator containing four oil capsules is located inside the fire box. (Photo by the author.)

In the previous statement, a single-acting or spring-return hydraulic actuator was unable to maintain a safety factor of 2 under all the conditions necessary for valve closure in the case study. As the double-acting actuator could produce higher torque values, the safety factor of 2 could be satisfied; therefore, a double-acting hydraulic actuator, rather than a single-acting hydraulic actuator, was selected for the third and last valve on the oil export line. In the case of a large-size ball valve, what are the implications of modifying the actuator selection? With a double-acting actuator, the ESD valve is failure-as-is and cannot achieve fail-safe closure (FSC). Thus, an accumulator with oil capsules was selected to stroke the valve to a fully closed position (Figure 8.8).

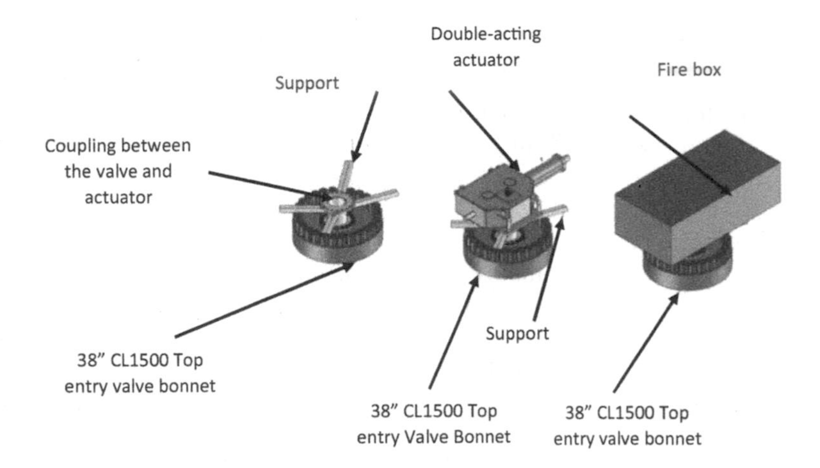

FIGURE 8.9 Fire box around the actuator as well as supports under the actuator. (Photo by the author.)

The oil bottles are pressurized by nitrogen to provide the force for three valve strokes. For safety reasons such as fire protection and explosion prevention, the capsules are kept in a fire box. There is a passive fire protection system, rigid type, designed to meet the most stringent protective requirements for the accumulator against hydrocarbon fires, pool fires, and jet fires.

Figure 8.9 illustrates how the fire box or passive fire protection must be installed around the double-acting actuator. As the double-acting hydraulic actuator is operated entirely with oil, it must be protected against an oil fire in the event of a fire. For the spring-return actuators, using a fire box is not a valid application, since the actuators are returned to a safe position, either open or closed, in the event of a fire. The fire box guarantees protection from pool fires up to 1100°C for a specific period until the double-acting actuator completely shuts off the pipeline valve in an ESD mode. A second consideration is to determine whether supports should be added to the coupling between the valve and actuator (see Figure 8.9) in order to increase the coupling's resistance against the actuator and fire box weights.

QUESTIONS & ANSWERS

1. Choose the incorrect or incomplete sentence about valves' actuators.

   A. Actuators for valves serve only one purpose: to provide enough force to open or close the connected valve.
   B. An actuator is a mechanical or electrical device mounted on top of an industrial valve and used to move, control, and operate the valve automatically.
   C. Power sources for valve actuators include electrical, pneumatic (compressed air), or hydraulic power.
   D. As a simple definition, a valve actuator is a box mounted on top of a valve that receives a signal or power supply (air or oil) to produce forces (linear motion) or torques (rotary motion) that control valve movement.

Answer) In all cases, the answers are correct and complete except for option A which is incomplete because valve actuators have several functions as follows:

- ➢ Moving the valve closure member to a suitable open or closed position
- ➢ Holding the valve closure member in the desired position
- ➢ Providing enough force or torque necessary to seat the closure member and meet the required shutdown leakage class
- ➢ Providing fully open or fully closed or as-is failure mode, and
- ➢ Providing a certain amount of closure member rotation with the correct speed.

2. Figure 8.10 illustrates a 38" CL1500 pipeline valve for installation on an oil export pipeline. Which of the following is true regarding the valve and its actuator?

FIGURE 8.10 A 38" CL1500 pipeline valve for an oil export pipeline. (Photo by the author.)

A. The valve has an emergency shutdown (ESD) function.
B. It is the first valve installed after the pig launcher.
C. The actuator on the top of the valve produces a linear motion for the valve operation.
D. The valve is actuated since the actuator provides ease of operation.

Answer) Because the valve has an electrical actuator, it cannot provide an ESD function; therefore, option A is incorrect. Since both the first and second valves after the pig launcher could be powered by electrical actuators, option B is incorrect. Option C is false since the ball valve requires a rotary motion for operation, rather than a linear one. The correct answer is D.

3. In Table 8.3, torque values are given for pipeline ball valves and their associated actuators. The project specifies that hydraulic and electrical actuators should have safety factors of at least 2 and 1.5, respectively. Which statement is correct?

TABLE 8.3
Torque Table for a Pipeline Valve

	Valve Torque Values (Nm)			
BTO	**RUN**	**ETO**	**BTC**	**ETC**
368939	59510	159794	358939	149794
	Actuator torque values (Nm)			
BTO	RUN	ETO	BTC	ETC
647878	123020	269588	687878	289588
	Safety factor values			
1.76	2.06	1.69	1.92	1.93

A. The valve's actuator is operated by hydraulic oil.
B. The actuator of the valve is a linear type.
C. For this valve, an electrical actuator is selected.
D. There is a fail-safe function for closing the valve's actuator.

Answer) The actuators of pipeline valves can be powered either by electricity or hydraulic oil. It is recommended that hydraulic actuators in this example provide torque values that satisfy a safety factor of at least 2. However, the actuator in this example does not meet the required safety factor except for run or running torque. Accordingly, a hydraulic actuator is not selected for the pipeline valve, and option A is incorrect. Option B is also incorrect since the pipeline valve is a ball valve requiring a scotch and yoke actuator and not a linear actuator. Option C is the correct answer. Option D is incorrect because the electrical actuator has a fail-as-is function.

4. Which type of actuator is suitable for the second pipeline valve after the pig launcher if it is a through conduit gate valve?

A. Linear electrical actuators
B. Double-acting scotch and yoke hydraulic actuator
C. Rotary electrical actuators
D. Single-acting scotch and yoke hydraulic actuator

Answer) Option A, the linear electrical actuator, is the correct answer.

5. To allow the actuator to move to a fail-safe closed position, which component or facility is used with the double-acting hydraulic scotch and yoke actuator?

   A. Coupling between the valve and actuator
   B. Accumulator
   C. Fire box
   D. None of above

   Answer) Option B is the correct answer.

BIBLIOGRAPHY

1. Equinor. (2019). Johan Sverdrup. [Online]. Available from: https://www.equinor.com/en/what-we-do/johan-sverdrup.html [access date: 6th January 2022].
2. Flowserve. (2018). Valve actuation: The when, how, and why of actuator selection. [Online]. Available from: https://www.flowserve.com/sites/default/files/2018-05/(FLS-VA-EWP-00005-EN-EX-US-0518-Actuation_Type_Advantages_LR1.pdf [access date: 6th January 2022].
3. Indelac Controls, Inc. (2014). How to select an actuator. Comprehensive Guide. [Online]. Available from: http://www.indelac.com/pdfs/How-to-Select-an-Actuator-Comprehensive-Guide.pdf [access date: 6th January 2022].
4. Sotoodeh K. (2015). Top entry export line valve design considerations. *Valve World Magazine*, 20(5), 55–61.
5. Sotoodeh K. (2019). Actuator sizing and selection. *Springer Nature Applied Science, Springer Switzerland*, 1, 1207. https://doi.org/10.1007/s42452-019-1248-z
6. Sotoodeh K. (2021). *Prevention of actuator emissions in the oil and gas industry*, (1st edition.). Austin, USA: Elsevier (Gulf Professional Publishing).
7. Sotoodeh K. (2021). Safety and reliability improvements of valves and actuators for the offshore oil and gas industry through optimized design. University of Stavanger. PhD thesis UiS, no. 573.
8. Sotoodeh K. (2022). *Coating application for piping, valves and actuators in offshore oil and gas industry*, (1st edition). UK: CRC Press.

9 Testing

9.1 INTRODUCTION

Valve testing is an essential part of the valve life cycle in the oil and gas industry. Testing valves after their design, manufacture, and assembly is essential for detecting potential valve defects. Pipeline valves on offshore platforms are subjected to a variety of loads for a very long period of time (e.g., 20–30 years) and operate under high pressure. It is critical for valves, especially those installed on offshore pipelines, to operate correctly and without any failures or external leaks throughout their operational life. The failure of a valve during operation could result from the failure or corrosion of soft or metallic materials, internal leakage, improper design and manufacturing, poor assembly, etc. When valves fail, they must be maintained or replaced. The maintenance and replacement of valves during operation can be extremely costly due to the possibility of process or plant shutdowns, loss of assets, and the need to replace a part or the entire valve. A valve failure and the subsequent leakage of a hazardous, highly pressurized, and potentially flammable fluid or gas outside the valve pose significant health, safety, and environmental (HSE) risks, including environmental pollution and harm to onsite personnel. Figure 9.1 illustrates the benefits of testing pipeline valves.

A valve should be internally tested by the valve manufacturer and have a test record before an external inspector (e.g., end user or purchaser) performs the ***factory acceptance test*** (***FAT***) and witnesses the pressure test. Pipeline valves are considered critical valves and should all be tested both internally and during the FAT. The FAT is performed on valves, including pipeline valves, following their design and assembly. The FAT is the final set of tests (e.g., pressure and function tests) conducted by the valve manufacturer at the factory to ensure that it complies with the order and project specifications. FAT is typically conducted in the presence of an external

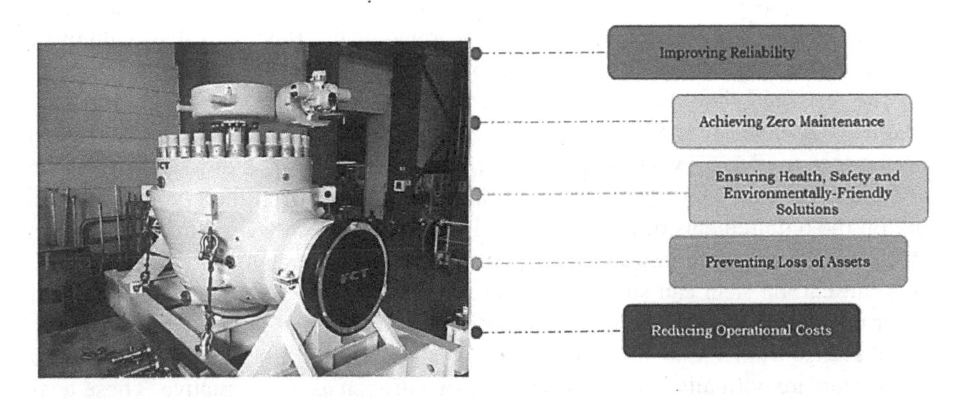

FIGURE 9.1 The benefits of testing pipeline valves. (Photo by the author.)

DOI: 10.1201/9781003343318-9

inspector and/or client. ***Production testing*** refers to the internal testing of the valves after production and assembly but before the final acceptance test.

9.2 TEST PROCEDURE

The first step in valve testing, including pipeline valves, is to establish a test procedure that will effectively test the valve. In general, valve manufacturers design their test procedures according to international standards, their own knowledge and experience, and client requirements. These are some of the main internal standards used for testing pipeline valves:

- American Petroleum Institute (API) 6D/ International Organization for Standardization (ISO) 14313, Specification for pipeline valves
- The American Society of Mechanical Engineers (ASME) B16.34, Specification for flanged, threaded, and buttwelding end valves
- API 598, Valve inspection and testing

9.3 TEST PREPARATION

Tests are conducted to ensure the correct assembly and performance of the valve in terms of tightness. All tests should be conducted at the valve manufacturer's facility. The test fluid pressure measurement equipment (i.e., pressure gauge) should measure the fluid pressure within an accuracy of ±5% of the required test pressure. The accuracy of the pressure gauge refers to the difference between the true and indicated read values on the gauge. For example, if a valve is pressure tested with 375 bar, the maximum difference between the read value from the gauge and the actual value is 5% of 375 bar or 18.75 bar. Furthermore, care should be taken to ensure that the test pressure does not fall below 25% or exceed 75% of the maximum pressure indicated on the pressure gauge. A pressure gauge should be calibrated periodically (e.g., on a six-monthly basis) to ensure that it has the required and desired level of accuracy. Some test benches are automatic and have controls on the console. There are digital pressure gauges on the control panel which demonstrate very high accuracy for a specified period of time before they are calibrated again. Pipeline valves cannot be tested on test benches because they are too large to fit on a bench.

The test fluid should be specified clearly in the test procedure. It can either be liquid or gaseous. In a fluid test, water containing some percentage of corrosion inhibitor is used (e.g., water plus 3% corrosion inhibitor). In addition, the chloride content of the water during the test can be at most 30 parts per million (ppm), depending on the requirements of the purchaser. In tests of pipeline valves made of carbon steel, the chloride limit in water is critical. In contact with a high amount of chloride content, carbon steel can suffer from pitting corrosion, so limiting the amounts of chloride in the test water may prevent pitting corrosion. The test water temperature should range from 5°C to 40°C. Several valve tests, such as a low-pressure test on the valve seat, are normally performed with air or nitrogen as an alternative. These tests are known as ***pneumatic tests***. For high-pressure gas tests, other gases such as inert gases (e.g., helium), nitrogen, or a mixture of helium and nitrogen may be used. Test pressures, except in low-pressure seat testing, are determined by the nominal pressure (PN) of the valves in accordance with international standards such as ASME B16.34. Typical PNs for pipeline valves include CL600 (PN100, 100 bar), CL900 (PN150, 150 bar), CL1500 (PN250, 250 bar), and CL2500 (PN420, 420 bar). The piping and valves in CL600 and higher are generally considered high pressure.

9.4 TEST SETUP

Testing pipeline valves can be potentially hazardous due to the high pressure of the test fluid or gas; therefore, any personnel involved in such testing should be aware of this danger and receive appropriate training to perform the test safely. Safety measures must be taken to ensure the safety of those performing the test and those who attend the testing for a variety of purposes, such as inspection. Therefore, it is proposed that the test area be shielded by barriers, or that the test is conducted inside a pit underground (see Figure 9.2), and that the test area be equipped with a camera for observing the test (see Figure 9.3). Pipeline valves are large in size, and thus cannot be fit on a test bench, so they must be tested outside of a test bench. Valve surfaces should not be painted or coated during the shell or body pressure test in order to avoid concealing possible leaks. In addition to this, valves must be thoroughly cleaned, dried, and free of dust, grease, or other lubricants that may be used during valve assembly. Bolts need to be tightened to the torque specified in the valve documentation provided by the valve manufacturer, such as the ***Installation, Operation, and Maintenance (IOM)*** manual. IOMs are documents provided by valve manufacturers that outline the critical points and procedures regarding valve installation, operation, and maintenance.

FIGURE 9.2 Testing a pipeline valve inside a pit (top view). (Photo by the author.)

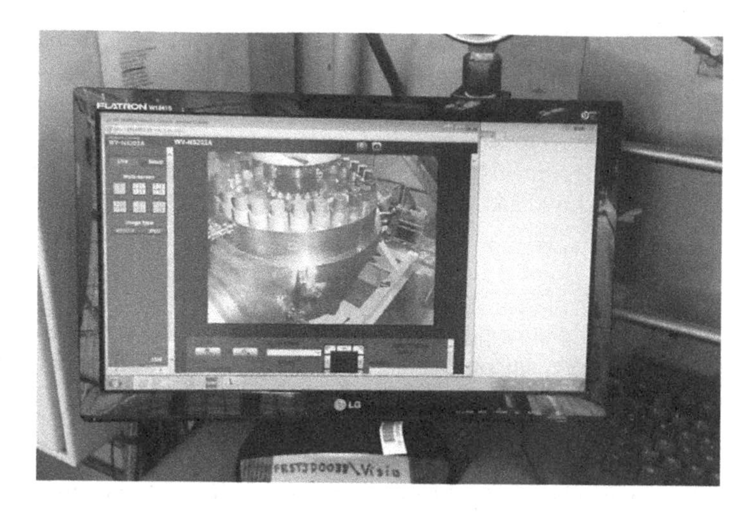

FIGURE 9.3 Using a computer and camera to demonstrate a valve in a pit during the test. (Photo by the author.)

FIGURE 9.4 Blinding a pipeline valve end with a cap for pressure test. (Photo by the author.)

Both ends of the pipeline valves are blinded with a couple of blind hubs, flanges, or caps to isolate the pressure source during the pressure tests. A pipeline valve end cap is shown in Figure 9.4 in order to cover the end of the valve during the pressure test. As a disadvantage, the cap must be cut after the pressure test, and the pup piece requires re-beveling to be ready for welding in the construction yard. On the other hand, a blind hub (see Figure 9.5) that is connected to the pup piece by a couple of clamps is not welded, which means the pup piece does not need to be re-beveled after disassembling the blind hub and clamps. If a pressure relief valve (PRV) has been installed on the body of the pipeline valve under test, as it may be the case if the valve is equipped with a double isolation and bleed 1 (DIB1) function, it should be removed before the pressure test and its connection should be plugged. The DIB concept is discussed in greater detail in Chapter 3.

FIGURE 9.5 Blind hub and clamp connections are used to blind the end of a pipeline valve during pressure testing. (Photo by the author.)

9.5 TYPE OF TESTS

Following tests should be conducted on every pipeline valve:

1. Hydrostatic high-pressure body or shell test
2. Hydrostatic high-pressure seat or closure tests
3. Hydrostatic functional and torque measurement tests
4. Hydrostatic cavity tests
5. Backseat test that applies to through conduit gate (TCG) valves
6. Air or gas low-pressure seat or closure tests
7. High-pressure gas body or shell test
8. High-pressure gas seat or closure tests
9. Drift test
10. Electrical continuity or antistatic test
11. Fire test

The following section aims to define the correct actions and required operations for the performance of the tests.

9.5.1 Hydrostatic High-Pressure Body or Shell Test

Tests on the valve body pieces or shell are intended to evaluate the structural integrity and robustness of the parts as the main pressure-containing parts against the internal fluid pressure. During a body pressure test, the packing and seals, as well as the plugs connecting the valves, are examined. Pressure-containment parts of valves include those that can lead to the release of internal fluid into the environment in the event of a failure. During a high-pressure body or shell test, the pipeline valve, whether a ball or a TCG, must be in a half-open position. In addition, the external surfaces of the valve shall be thoroughly dried in order to detect possible leaks from the valve more easily. The valve shall be completely filled with water prior to testing. The operator must ensure that the valve is completely deaerated. Valve body and bonnet are subjected to a hydrostatic pressure that is 1.5 times the valve's nominal pressure (PN) at ambient temperature. Accordingly, if the valve can withstand a pressure value of 259 bar at ambient temperature, then the hydrostatic high-pressure body test applied pressure is 388 bar. In Tables 9.1 and 9.2, the hydrostatic shell and seat test values for carbon steel and low-temperature carbon steel (LTCS) materials that are preferred for the body and bonnet of pipeline valves are given. The operator will apply the test pressure through a port connected to one end of the valve. After stabilizing the pressure, the operator isolates the source of pressure. The operator records the pressure and writes this on the test certificate. Test duration depends on the size of the tested valve and is determined according to API 6D. Table 9.3 provides the minimum hydrostatic, pneumatic, and gas shell or body test durations based on the size of the valve. It is recommended that the minimum hydrostatic shell test duration for a 38"

TABLES 9.1

Hydrostatic Shell and Seat Test Values for Carbon Steel and Low-Temperature Carbon Steel (LTCS)

Standard Class Materials

A216 Gr.WCC
A352 Gr.LC2
A352 Gr.LC3
A352 Gr.LCC

CL 150 (bar)	CL 300 (bar)	CL 400 (bar)	CL 600 (bar)	CL 900 (bar)	CL 1500 (bar)	CL 2500 (bar)
			Maximum service pressure			
20	52	69	103	155	259	431
			Shell test pressure			
30	78	104	156	233	388	647
			Seat test pressure			
22	57	76	114	171	285	474

TABLES 9.2

Hydrostatic Shell and Seat Test Values for Carbon Steel and Low-Temperature Carbon Steel (LTCS)

Standard Class Materials

A105
A216 WCB
A350 LF2

CL 150 (bar)	CL 300 (bar)	CL 400 (bar)	CL 600 (bar)	CL 900 (bar)	CL 1500 (bar)	CL 2500 (bar)
			Maximum service pressure			
20	51	68	102	153	255	425
			Shell test pressure			
30	77	103	154	230	384	639
			Seat test pressure			
22	57	76	113	169	281	468

TABLE 9.3
Minimum Duration for Hydrostatic, Pneumatic, and Gas Shell Tests as Defined by API 6D

	Minimum Shell Test Duration as Per API 6D (Minute)	
	Hydrostatic Shell Test	Pneumatic or Gas Shell Test
≤4"	2	15
6"–10"	5	15
12"–18"	15	15
≥20"	30	30

ball or TCG valve is 30 minutes; Table 9.3. During the shell test, the operator inspects the external surfaces of the valve after the stabilization time and during the test, and no visible external leaks are permitted. The presence of humidity on the external valve surface and the drop at the pressurizing test port indicate leakage from the valve and may result in valve rejection. During the high-pressure hydrostatic shell or body test, the tightness of sealant injection and flushing ports should be checked. For pipeline valves equipped with sealant injection ports on seats and stems, all the external nipples used to extend sealant injection ports should be removed, and the tightness of the mini check valves on sealant injection ports should be checked for no leakage during the test.

9.5.1.1 Safety Relief Valve Test

This test applies to DIB1 valves which require a safety relief valve to automatically release excessive pressure from the valve body cavity in the event of an overpressure condition. As previously explained, any PRV installed on the tested valve should be removed before the test begins. As soon as the hydrostatic shell test is completed, the PRV should be reinstalled, and the valve should be in the half-open position. By means of the port connected to the valve body, the operator should pressurize the valve by 95% of the safety relief valve set pressure. At the same time, pressure is increased within the cavity until the safety valve operates and relieves the pressure. It is the operator's responsibility to record the relief pressure in the test report, which should be between 1.1 and 1.33 times the maximum service pressure. The set pressure, which can be expressed in bar or pounds per square inch (psi), is the pressure at which a safety or PRV opens.

9.5.2 Hydrostatic High-Pressure Seat or Closure Tests

As part of a seat test, also known as a leakage test, the seal between the closure member and the seats of the valve is evaluated. Before performing this test, it should be ensured that any lubricants used to assemble the seats and closure member are removed from them and from the sealing surfaces of the seats. The operator should move the valve from the partially open position to the closed position and apply a test pressure equal to 1.1 times the PN as indicated in Table 9.1 or Table 9.2 to one seat. In accordance with API 6D, the other side of the tested seat is at atmosphere pressure. The minimum duration during which the pressure should be maintained is given in Table 9.4. Any leakage from the seat flows into the valve cavity, so it is imperative to inspect the cavity vent and drain connections for any possible leakage. Figure 9.6 illustrates a hydrostatic leakage detector that can count leaks during testing.

TABLE 9.4
Minimum Hydrostatic, Pneumatic, and Gas Seat or Closure Test Duration as Defined in API 6D

Minimum Seat or Closure Test Duration as per API 6D (Minute)		
	Hydrostatic Seat Test	Pneumatic or Gas Seat Test
≤4"	2	15
6"–18"	5	15
≥20"	10	30

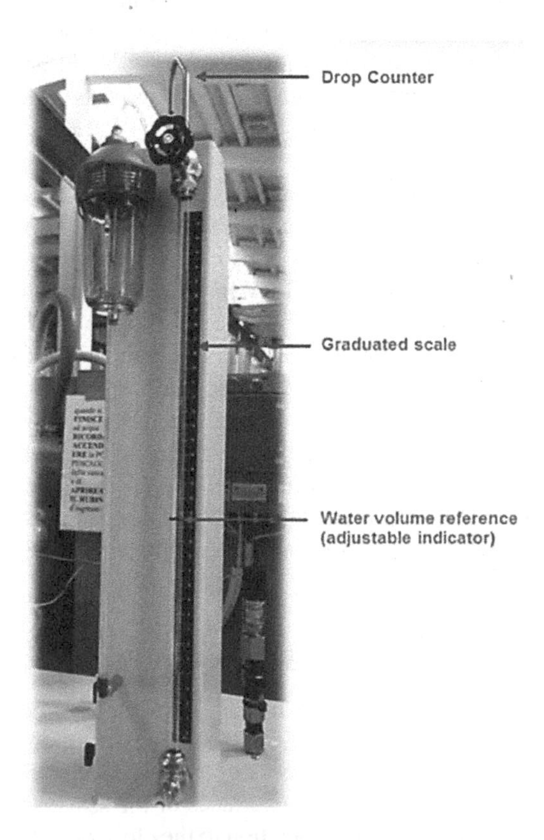

FIGURE 9.6 A hydrostatic leakage detector. (Photo by the author.)

No leakage is allowed from soft or non-metallic seats, but some leakage is allowed from metallic seats. ISO 5208 and EN 12266 specify different leakage rates like A, B, C, D, E, F, and G. ISO 5208 specifies some additional leakage rates such as AA, CC, and EE. Leakage rate A implies no visually detectable leakage for the duration of the test. The amount of allowable leakage increases in alphabetical order of the letters used to indicate the leakage rates; for example, the letter G indicates a higher allowable leakage rate than the letter F. Furthermore, the notation CC indicates a higher allowable leakage rate than the letter C but a lower one than the letter D. The maximum allowable leakage rates are provided in units of cubic millimeter per second for both gas and liquid closure tests, drop per second for liquid closure test, and bubble per second for gas closure test. The maximum allowable leakage rates, except for leakage rate A, are dependent on the test fluid (gas or liquid) and the dimension nominal (DN) of the valve. For example, the maximum allowable leakage rate B for a liquid closure test is $0.01 \times \text{DN}$ in $\frac{mm^3}{s}$ or $0.00016 \times \text{DN}$ $\frac{drop}{s}$. Table 9.5 shows the maximum allowable closure test leakage rates per ISO 5208. Pipeline valves are metal seated due to reasons explained in Chapter 1, and their accepted leakage rates are usually given in the valve datasheets. The author experiences that end users specified leak rate B as per ISO 5208 during the high-pressure hydrostatic closure or seat tests for pipeline valves in a couple of offshore projects. Thus, leakage values must not exceed those mentioned in Table 9.5. Both ball and TCG valves have two seats, and hydrostatic high-pressure seat or closure test should be repeated for the other one.

9.5.3 Hydrostatic Functional and Torque Measurement Tests

By closing the valve completely and pressurizing the seat from the line, torque measurement tests can be performed on every seat. The test pressure is equal to the valve's PN. The operator should pressurize the first seat, or "seat A," open the valve, and record the torque required to move the valve from a closed to an open position. The test is repeated on the second seat or "seat B." The maximum measured opening and closing torques shall not exceed the torque values specified in the valve datasheets. In order to determine the opening and closing torque values, both seats could be under pressure from the line with the test pressure of the valve's PN. As part of the function test, it is also essential to determine the opening and closing times. The closing time of the ***emergency shutdown*** (ESD) pipeline valve is crucial, and during the function test, the measured closing time shall not exceed the value specified in the datasheet for the valve. Electrical actuators are commonly used for ease of operation, and they cannot operate valves at very high speeds. Therefore, it is not essential for electrically actuated valves to operate at a high speed.

9.5.4 Hydrostatic Cavity Tests

Valve cavity tests are applicable to valves that have a cavity in the body, such as ball valves and TCG valves. The cavity test, which is also known as the ***cavity relief test***, is a type of test that is conducted on ***self-relieving (SR)*** seats to ensure that the SR seat is able to relieve the excess pressure within the cavity to the inside of the valve. A pressure higher than 1.33 times the valve nominal or rated pressure is not

TABLE 9.5
Maximum Leakage Rates Allowed During Closure Tests According to ISO 5208

Test Fluid	Unit Leakage Rates	Rate A	Rate AA	Rate B	Rate C	Rate CC	Rate D	Rate E	Rate EE	Rate F	Rate G
Liquid	$\frac{mm^3}{s}$	No leakage	0.006 x DN	0.01 x DN	0.03 x DN	0.08 x DN	0.1 x DN	0.3 x DN	0.39 x DN	1 x DN	2 x DN
	$\frac{drops}{s}$		0.0001 x DN	0.00016 x DN	0.0005 x DN	0.0013 x DN	0.0016 x DN	0.0048 x DN	0.0062 x DN	0.016 x DN	0.032 x DN
Gas	$\frac{mm^3}{s}$	No leakage	0.18 x DN	0.3 x DN	3 x DN	22.3 x DN	30 x DN	300 x DN	470 x DN	3000 x DN	6000 x DN
	$\frac{bubbles}{s}$		0.003 x DN	0.0046 x DN	0.0458 x DN	0.3407 x DN	0.4584 x DN	4.5837 x DN	7.1293 x DN	45.837 x DN	91.673 x DN

(applicable for tests with liquid/air and gas)

permitted in the valve cavity. According to Figure 9.7, an SR seat design allows cavity overpressure to be released to the inside of the valve.

9.5.4.1 Both Seats Are SR

During a cavity test, the valve cavity must be pressurized; therefore, plugs or flanges are usually required as ports to pressurize the valve cavity. It is possible for three conditions to arise during the cavity relief test. First of all, both valve seats are self-relief types, which are relatively uncommon for pipeline valves. The valve cavity is pressurized with water in this case, and the pressure on both sides of the closure member of the valve is monitored for any possible leakage. During the cavity test, the valve is closed. As low as 1 bar of pressure can be applied to the cavity. This pressure should be gradually increased until leakage occurs from one side of the valve closure to the other side. The cavity pressure should be recorded during the release process. If the recorded cavity pressure during leakage from the seat is less than 133% × PN of the valve, the seat which relieves pressure from the cavity passes the cavity release test. The pressure should be increased since the other seat releases the cavity pressure to the inside of the valve in the range provided previously. Figure 9.8 illustrates a ball

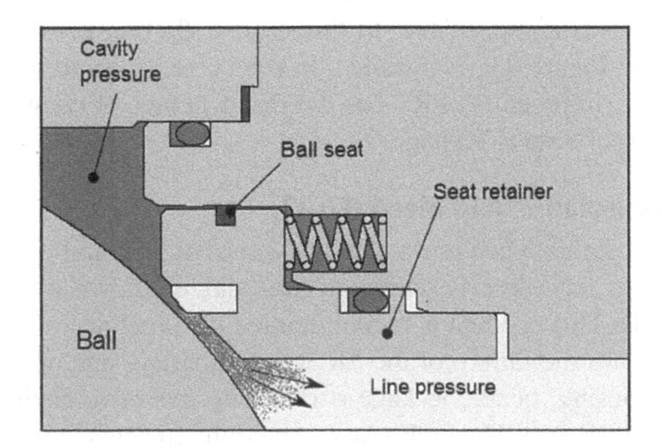

FIGURE 9.7 An SR seat for releasing excess pressure of the cavity into the valve interior. (Credited to the author.)

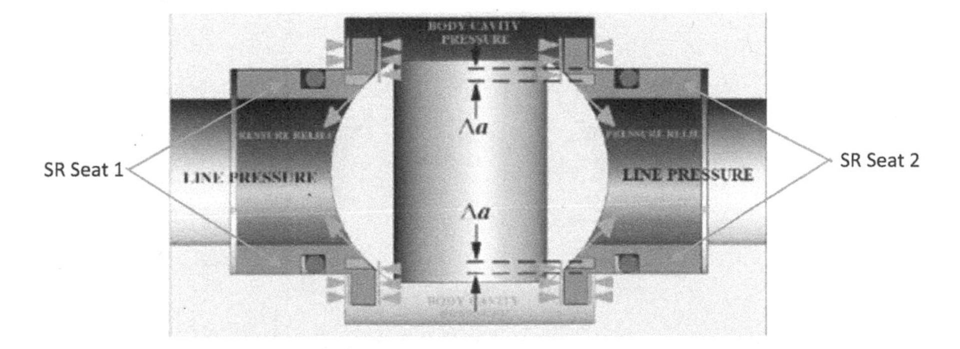

FIGURE 9.8 Ball valve with two SR seats that allow the pressure in the cavity to be released to the inside of the valve. (Photo by the author.)

valve with two SR seats that release the extra cavity pressure from both sides of the valve to the interior.

9.5.4.2 Double Isolation and Bleed (DIB1)

The second condition occurs when both seats are ***double piston effect (DPE)*** seats. In accordance with API 6D, a valve with two DPE seats is called a ***DIB1*** valve. As opposed to SR seats, DPE seats are unable to release the cavity pressure. A DPE seat is used to prevent or minimize leakage from the cavity to the inside of the valve. In Figure 9.9, we illustrate a ball valve with two DPE seats and a pressurized valve cavity. According to this figure, there is no leakage from the valve cavity into the valve from the DPE seats.

The valve is in the closed position during the test, and the cavity is pressurized with 1.1 times the valve PN with water, and the leakage on both sides of the closure member is measured. The amount of leakage from the cavity through the metallic seats is the same as the allowable leakage rate from the seats during hydrostatic high-pressure closure tests (e.g., ISO 5208 leak rate B). In a DIB1 valve with two DPE seats, one of the issues is the release of cavity pressure from the cavity. If the trapped fluid in the cavity is a gas, a pressure relief mechanism from the cavity is not necessary because the gas is compressible. In this case, a ***PRV*** or ***pressure safety valve (PSV)***, as shown in Figure 9.9, is installed on the valve cavity to release the excess pressure into the environment. A PRV has the disadvantage of creating a hole in the body cavity which can lead to leakage.

9.5.4.3 Double Isolation and Bleed (DIB2)

Another condition occurs when one valve seat is an SR seat and the other is a DPE seat, in which case the valve is unidirectional and is known as DIB2 in accordance with API 6D. This concept is recommended for pipeline valves in Chapter 3. Figure 9.10 illustrates the failure of the SR seat on the left side of the valve to seal the valve during closing; hence, leakage occurs into the valve cavity from the line passing through the SR seat. The DPE seat on the right prevents leakage of fluid from the cavity to the downstream side of the valve. Thus, the DPE seat can protect the operator who is performing maintenance on the downstream side of the valve.

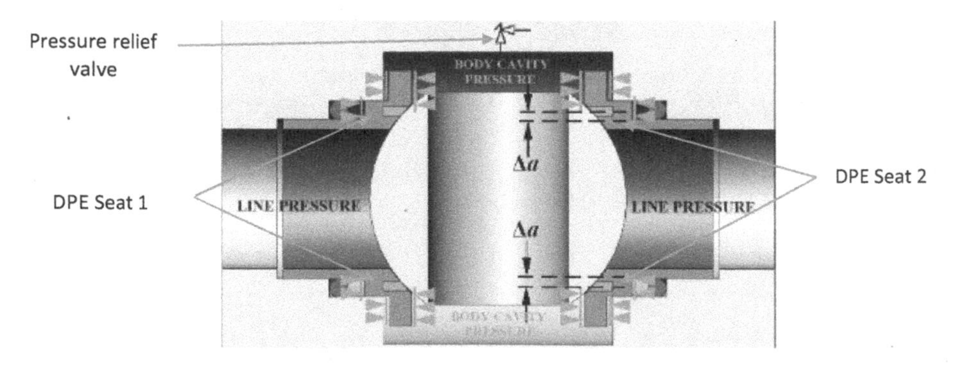

FIGURE 9.9 A ball valve with two DPE seats and a pressurized cavity. (Credited to the author.)

In order to perform a cavity test on a DIB2 valve, as in the other two cases, the valve must be closed during the cavity test. Second, two cavity tests are conducted: one for the DPE seat and one for the SR seat. In order to perform the DPE cavity test for a DIB2 valve, it is necessary to pressurize the cavity with water to a pressure that is 1.1 times the PN of the valve and monitor the line on the DPE seat (downstream side, Figure 9.10) for leakage. The maximum leakage rate from the DPE seat is the same as the maximum leakage rate from the seat during the hydrostatic closure test (e.g., leak rate B). As a result of the SR seat, the cavity pressure can be released to the valve inside, potentially concealing the leakage from the DPE seat. Therefore, the SR seat should be pressurized by the value to 1.1 times the PN from the line or upstream side during the DPE seat cavity test (Figure 9.10). In an SR test of a DIB2 valve, the cavity pressure is gradually increased from 1 bar, and the leakage is monitored in the valve close to the SR seat (upstream side, as shown in Figure 9.10). Pressure release from the cavity through the SR seat into the inside of the valve (on the upstream side, Figure 9.10) should occur when the pressure inside the cavity does not exceed 1.33 times the PN.

9.5.5 Backseat Test

The backseat test is applied to TCG valves. Another type of pipeline valve is the ball valve, which does not have a backseat feature, so they are not subject to this test. Figure 9.11 illustrates a slab gate valve that is a type of TCG valve. The main components of the valve, including the backseat, are illustrated. A valve's stem is in an upward position, and the backseat sits on the inner surface of the bonnet and prevents fluid service from entering the stem seal area. Therefore, the backseat area can be considered a secondary stem seal. The test operator applies water from one end of the valve with a minimum pressure value of 1.1 times the valve's PN according to API 6D (see Tables 9.1 and 9.2 for PN values). For the backseat test, water is used at an ambient temperature of 38°C (100°F). The minimum duration of the backseat test according to API 6D is provided in Table 9.6. In order to seal the stem seals, also

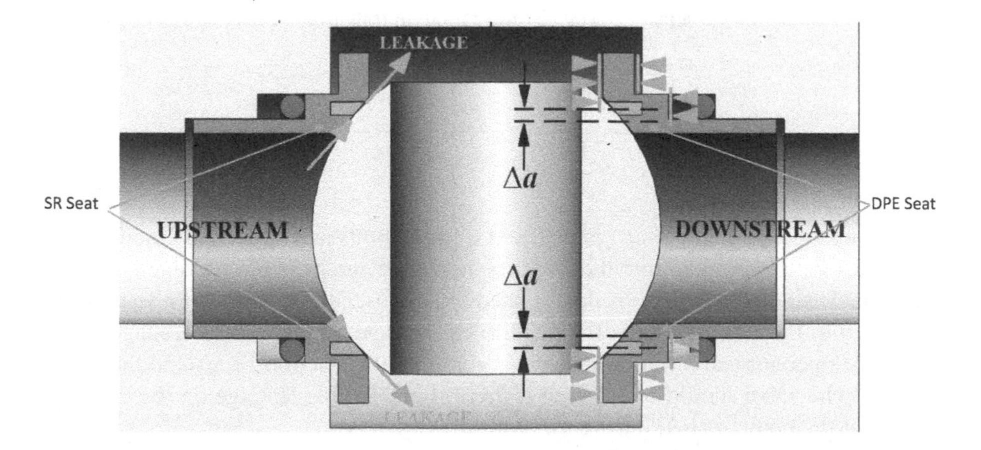

FIGURE 9.10 A DIB2 ball valve with one SR seat (left side) and one DPE seat (right side). (Credited to the author.)

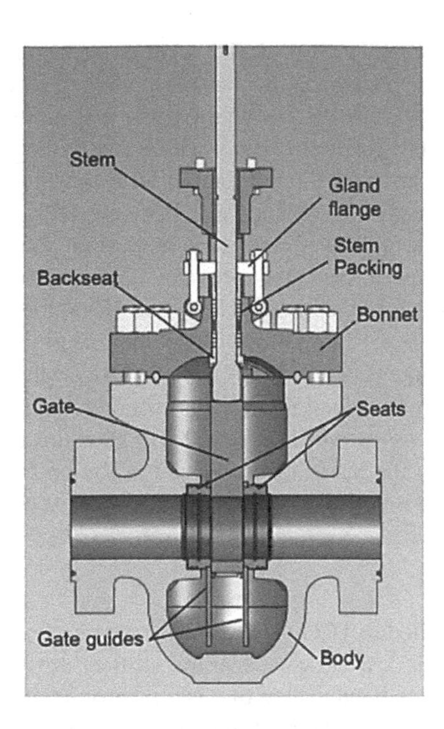

FIGURE 9.11 A slab gate valve and its parts.

TABLE 9.6
Minimum Duration for a Backseat Test Based on API 6D

NPS	DN	Test Duration (Minute)
≤4"	≤100	2
≥6"	≥150	5

called packing, the gland flange is fastened through bolts and nuts. The gland flange bolts must be loosened so that the operator is able to see possible leakage from the backseat. The gland flange bolts do not need to be loosened, however, if a test port is connected to the stem seal area for the purpose of detecting backseat leakage. As a result, the operator can monitor leakage from either the test port or around the loose packing. The valve should be rejected if there is any visible leakage on the external surface of the body or shell during the testing.

9.5.6 Air or Gas Low-Pressure Seat or Closure Tests

The test operator should drain and empty the valve from water before conducting the closure tests with air or gas. Three differences distinguish these tests from high-pressure closure or seat tests with water: the test fluid is air or gas (e.g., nitrogen), the maximum leakage is different, and the air or gas test pressure is different. API 598 specifies a low-pressure air or gas test as a value between 4 and 7 bars. The minimum closure test duration for the low-pressure air or gas test is shown in Table 9.4. The maximum allowed air or gas leakage during this test depends on the valve's leakage class or leakage rate (e.g., leak rate B or C). The leakage values are given in units of bubbles per second or cubic millimeters per second in Table 9.5. It is possible to detect possible leaks during an air or gas test using one of the following methods:

- Observing loss of bubbles via a bubble counter connected to the test port, as illustrated in Figure 9.12
- Monitoring pressure drops from the pressure gauge

9.5.7 High-Pressure Gas Body or Shell Test

Performing a high-pressure gas test is extremely hazardous, and the personnel involved, such as operators and inspectors, shall take appropriate safety measures

FIGURE 9.12 A bubble counter for detecting leaks of air or gas. (Photo by the author.)

during this process. Some end users, such as Equinor, may avoid performing high-pressure gas body or shell tests for pipeline valves because such testing is costly, time-consuming, and requires high safety precautions. It is possible to use an inert gas or a mixture of 99% nitrogen and 1% helium as the test medium. The minimum test pressure shall be 1.1 times the PN of the valve being tested. The operator will apply the test pressure through a port connected to one end of the valve. API 6D recommends submerging the valve in a water bath during this test in order to better detect possible leaks, which could appear as bubbles in the water bath and to enhance safety. During the high-pressure gas body testing, no external leakage is permitted. When the valve is tested in a water bath, observation of a bubble in the bath results in the valve being rejected during the high-pressure gas shell test. Table 9.3 indicates the minimum gas shell test duration.

9.5.8 High-Pressure Gas Seat or Closure Test

The test medium may consist of an inert gas, nitrogen, or a mixture of 99% nitrogen and 1% helium. However, API 6D emphasizes the use of inert gas as a test medium. There are two main differences between these tests and the high-pressure closure or seat test with water: the test fluid is gas (e.g., nitrogen), and the maximum leakage limit is different. The minimum test pressure shall be 1.1 times the PN of the valve at room temperature (38°C), and the minimum test duration should be based on Table 9.4. API 6D does not permit leakage rates exceeding ISO 5208, rate D. However, the end user can specify a more stringent leakage rate than API 6D, which takes precedence over the standard requirement. As an example, the author encountered an end user who requested ISO 5208 leak rate C for pipeline valves during a recent offshore project involving a high-pressure gas seat or closure test. In addition, it was noted by the author that the test operator repeated the high-pressure gas seat or closure test on each seat ten times after opening and closing the valve against full differential pressure to measure the seal ability of valve seats after cycling (opening and closing).

9.5.9 Drift Test

Since pipeline valves are subject to pigging, they must undergo a drift test. In Chapter 1, pig stands for pipeline inspection gadget that is shot inside the pipeline and within the valves through a pig launcher located before or upstream of the valves. This is for cleaning, inspection, and maintenance purposes. A mandrill of the same dimensions as the pig should pass completely through the valve bore without any obstruction during drift test. Figure 9.13 illustrates a drift test for a ball valve installed on a gas export line with a 280-mm bore. The mandrill consists of three disks of polytetrafluoroethylene (PTFE) with a diameter of 280 mm, which corresponds to the outside diameter of the valve bore. The first disk is located in front, highlighted in Figure 9.13; the second disk is located in the middle, inside the obturator bore area; and the third disk is located at the backside of the valve. The mandrill should pass through the valve in both directions. It is not an expensive nor time-consuming test, and it is limited to valves that can be piggable, such as pipeline valves.

FIGURE 9.13 A pipeline ball valve during drift test. (Photo by the author.)

9.5.10 Electrical Continuity Test

In addition to the antistatic test, this test is also known as the electrical resistance test, which measures the electrical resistance between the closure member, valve's body, and stem or shaft when the valve is connected to a 12-volt power source. The electrical resistance measured on the selected path between these three valve components must not exceed 10 Ω. An antistatic test is performed to ensure that a valve is fire-safe. As a result of electrical resistance exceeding 10 Ω, static electricity can accumulate in the valve closure member, body, and stem assembly, posing a risk of fire inside the valve.

9.5.11 Fire Test

The purpose of a fire test is to ensure that the valves provide the required sealing and function according to the fire test standard and to confirm the design is fire-safe. There are different standards for the performance of fire tests on valves, including API 6FA (specification for fire testing of valves), API 607 (fire testing for quarter-turn valves and valves equipped with non-metallic seats), and ISO 10497 (fire testing requirements for valves). Every standard has its own test method. The test methods discussed in this section are based on API 6FA since this standard is very common.

It is necessary to examine two related issues before discussing fire test methods. First of all, the fire test of valves is potentially hazardous, and therefore, the safety of those performing the test should be the primary concern. In view of the nature of the test and possible issues with the design of the test valve, hazardous ruptures may occur at the valve pressure boundaries. Hence, the personnel should be protected by providing an enclosure around the fire test area. Second, a fire test is a kind of type test and not a production test. Suppose, for example, that the valve manufacturer does not have a fire test certificate for a specific type of valve ordered by the purchaser. As a result, the valve manufacturer typically manufactures one extra valve to perform a fire test and validate the fire-safe design of the new valve. In contrast, if the valve manufacturer has fire test certificates for all ordered valves, a fire test is not required for the valves included in the order.

9.5.11.1 Fire Test According to API 6FA

The valve should be filled with water and closed as the test medium; it is installed horizontally with a vertical stem. The fire temperature should range between 761 and 980°C, and the test should last 30 minutes. Two or three thermocouples are installed around the valve body, depending on the size of the valve being tested. A pipeline valve above 6" is usually equipped with three thermocouples (see Figure 9.14).

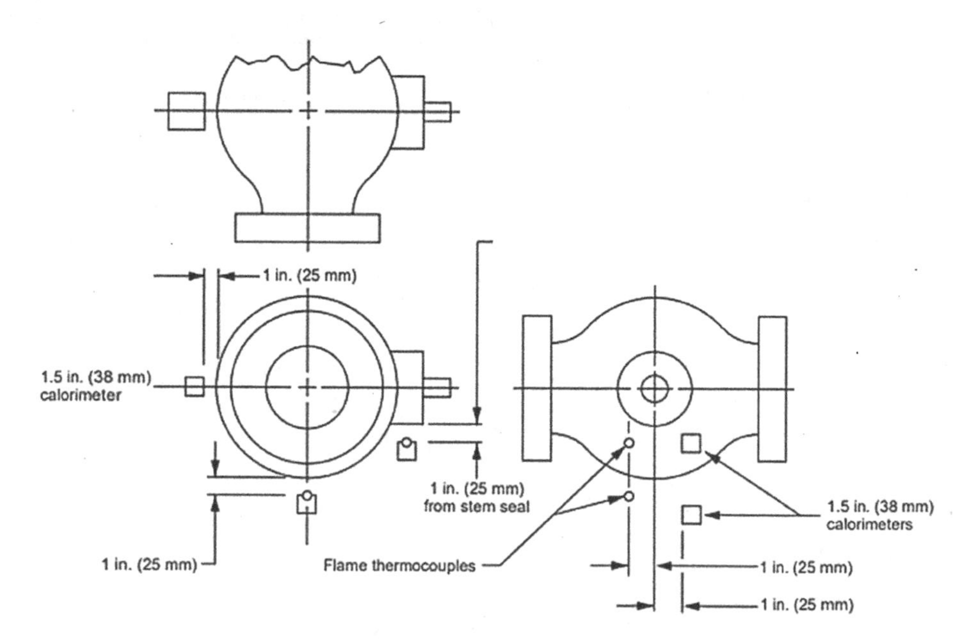

FIGURE 9.14 The location of thermocouples and calorimeters on large valves of 8" or larger during fire tests as per API 6FA.

The thermocouples are installed at the center of a 1.5" cubic inch ***calorimeter*** block made from carbon steel; a calorimeter measures heat.

During or after the fire, when the valve has cooled down, a maximum of seven tests should be conducted on the valve. Most of these tests involve measuring internal and external leakage. Also, various types of tests, such as operability and pressure release from the valve's body cavity, are included. They are as follows:

1. Through seat leakage (high-pressure test) – during burn period

   During 30 minutes of burning, the maximum leakage should not exceed 400 milliliters per inch of the valve per minute (ml/inch/min). API 6FA provides high-pressure test values for valves designed in accordance with API 6D based on the valve's PN. Table 9.7 provides high- and low-pressure test values according to API 6FA for valves designed according to API 6D.
2. Body external leakage (high-pressure test) – during burn and cool down period

   The maximum external leakage should not exceed 100 milliliters per inch of valve size per minute during the 30-minute test period followed by cooling down to 100°C.
3. Through seat leakage (low-pressure test) – after cool down

   The test is applicable to valves of pressure class 600 and below. During the first five minutes after a valve has cooled down, the leakage from the seat should not exceed 40 milliliters per inch of valve size per minute. When the valve temperature reaches 100°C, the cooling period is complete.
4. Body external leakage (low-pressure test) – after cool down (valve in closed position)

   The test is also applicable to valves of pressure classes 600 and below. It is recommended that the maximum external leakage from the valve during 5 minutes of testing after the valve has cooled down should not exceed 20 milliliters per inch of valve size per minute.

TABLE 9.7
High- and Low-Pressure Test Values as per API 6FA for Valves Designed as per API 6D

Valve Rating		High Test Pressure		Low Test Pressure	
Class	PN	psi	bar	psi	bar
150	20	210	14.5	29	2.0
300	50	540	37.2	50	3.4
400	64	720	5.0	70	4.8
600	110	1080	7.5	105	7.2
900	150	1620	11.2	-	-
1500	260	2700	18.6	-	-
2500	420	4500	31.0	-	-

5. Operation of test valve after fire test

   In order to operate properly, the valve should be able to be unseated from the closed position against a high-pressure differential and be able to move from that position to the open position.

6. Body external leakage in open position

   The body external leakage of the valve in an open position should not exceed 200 milliliters per inch of valve size per minute.

7. Pressure relief provision

   The SR seats should be tested to ensure they release the accumulated pressure in the cavity. The accumulation of pressure in the body cavity to more than 1.33 times the valve design pressure is not permitted, as this could damage the valve body and bonnet. The cavity pressure can be increased from 1 bar to a higher value step-by-step until the pressure gauge shows a reduction in the cavity pressure. At a pressure value not exceeding 1.33 times the valve design pressure, the cavity pressure should be released.

9.5.11.2 Fire Test for Pipeline Valves

Notably, some valve sizes and pressure classes may be qualified automatically without any fire testing requirement provided that a similar valve had been tested and qualified with the same design and non-metallic materials as the untested valve. One test may be used to qualify valves larger than the tested valve but not exceeding two times the size of the tested valve. The valve size of 16" qualifies all larger valve sizes, and one test valve may also qualify valves with higher pressure ratings, but not exceeding two times the pressure rating of the valve that is being tested. Pipeline valves have unique designs, sizes, and pressure classes that haven't been qualified before, so it's impossible to accept their design for the fire based on previously prepared valves. A valve manufacturer would find it very expensive to make such a large and expensive pipeline valve and burn it during the fire test. An alternative solution is to apply a thermal analysis to the pipeline valve by simulating fire test parameters such as temperature and duration using software such as ANSYS to analyze the valve behavior under fire conditions such as leakage and expansion. An engineer who works with the software models the valve from the valve's general arrangement drawing, which details the valves' dimensions.

9.5.11.3 Fire Thermal Simulation and Analysis Case Study

In this section, we review a thermal analysis case study for a 38" CL1500 pipeline ball valve to evaluate the valve's operability during a fire. On the outer surface of the valve, a temperature of 1100°C was applied (see Figure 9.15).

After 2 minutes and 30 seconds (150 seconds), the valve internals and body would remain unchanged without any expansion, according to the software analysis. In Figure 9.16, which was extracted from the software, we can see how 20°C is transferred to the valve body interior and to the valve internals after 150 seconds of operation. As a result, the internal parts of the valve, such as the ball and seat, will not experience any expansion or dimensional change during the fire before 150 seconds.

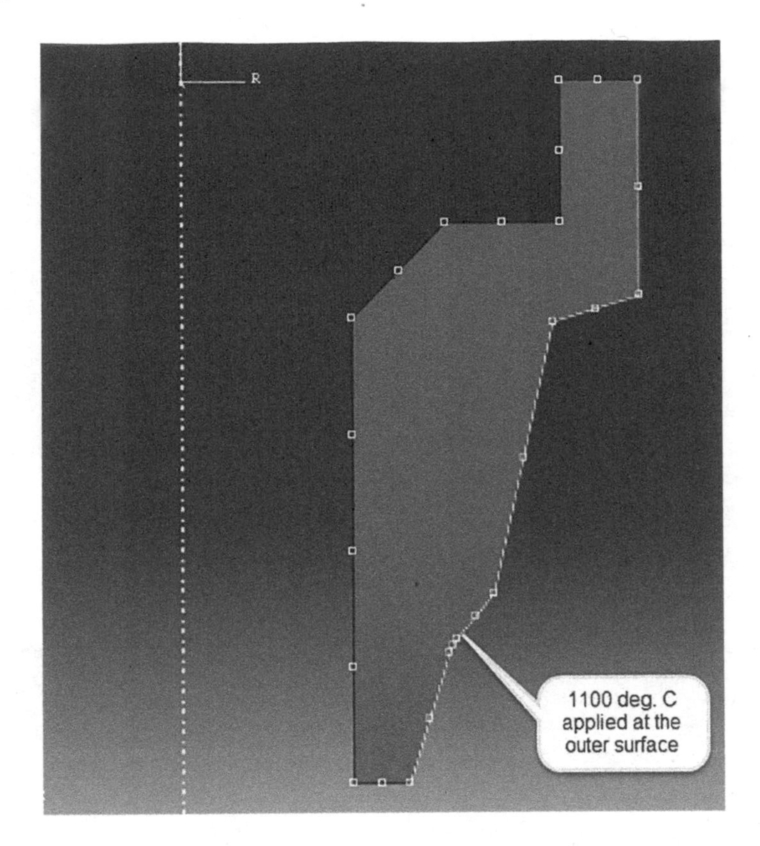

FIGURE 9.15 Applying 1100°C to the outer surface of the pipeline valve. (Photo by the author.)

This analysis examined the displacement of the body in contact with seat after a 150-second fire, as shown in Figure 9.17. An excessive expansion of the seat can prevent rotation of the ball or closure member during valve operation, thereby jeopardizing the valve's functionality. The maximum expansion of the body was 0.34 mm, which is minor enough not to affect the operability of the valve during a fire. In addition, the expansion of the valve body will not affect the position of the seat and the operability of the valve since there is an air gap between the expanded area of the body and the seat more than one millimeter, providing the required space for the body expansion without any friction with the seat. Minor body expansion and no changes in the internals of the valves indicate that in a fire, the ball of the valve could move freely inside the valve without any obstruction. This analysis was conducted to validate fire test standards for valve operability during a fire.

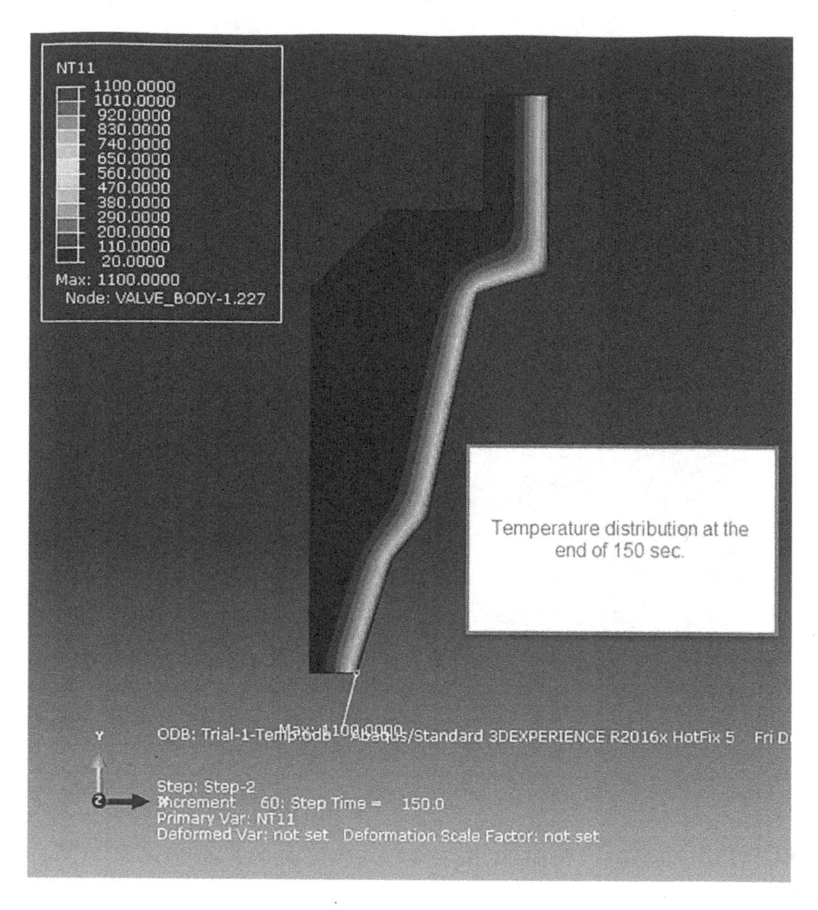

FIGURE 9.16 Temperature distribution on the valve body after 150 seconds. (Photo by the author.)

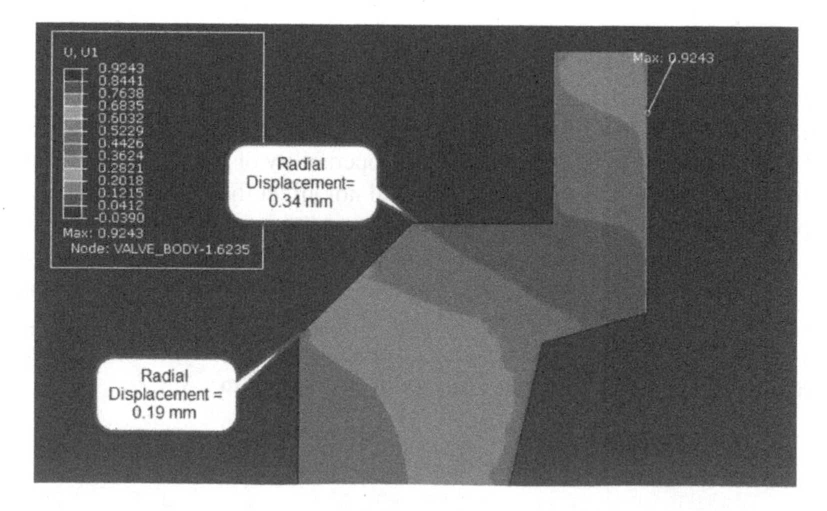

FIGURE 9.17 Maximum radial displacement in the body of the valve. (Photo by the author.)

9.6 ACTUATORS TESTING

As discussed in Chapter 8, an actuator is a machine or component mounted on the top of an industrial valve in order to move and control the valve automatically. It is the actuator that determines how well the valve performs. There are two types of actuators used for pipeline valves: electric and hydraulic.

9.6.1 Hydraulic Actuators Testing

The hydraulic actuators can be classified as single-acting (spring-return) actuators, with a spring on one side and a hydraulic connection (piston) on the other, or double-acting actuators, with hydraulic and piston connections on both sides (see Figure 9.18).

All hydraulic actuators must be pressure tested before being installed on valves. In order to verify the structural integrity of actuators, the operator tests each hydraulic cylinder and piston. Each hydraulic piston/cylinder should be tested with 1.5 times the actuator design pressure for 15 minutes. Typically, hydraulic actuators operate at a hydraulic pressure of 160–200 bar. The test pressure should be at least 1.5 times the hydraulic working pressure. Nevertheless, the test pressure can be increased to a minimum of 1.5 times the hydraulic design pressure, which is 1.1 times the hydraulic working pressure. For a hydraulic working pressure of 160 bar, the design pressure is 1.1 times 160, or 176 bar, and the shell test should be done at either 240 bar (1.5*160 = 240 bar) or 264 bar (1.5*176 = 264 bar). During the shell test, there should be no visible leakage from the actuators. Furthermore, any hydraulic pressure drop of more than 3% of the test pressure after one hour is considered a sign of hydraulic leakage and actuator rejection. Accordingly, the maximum pressure drop for the 15-minute shell test of a hydraulic actuator with 240 bar is 240 bar x 0.03 x 15/60 = 1.8 bar. The loss of hydraulic pressure below 238.2 bar is considered a leakage from the actuator. One important actuator test is known as the seal test, in which the actuator seal should be placed under pressure equal to 0.2 times the hydraulic working pressure. The test could last for three minutes. It is prohibited to allow leaks of any kind into the spring chamber from the actuator seals, such as the piston

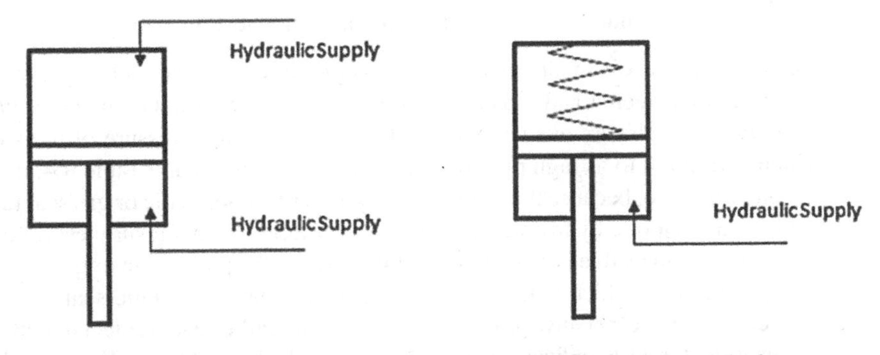

FIGURE 9.18 Double-acting and spring-return actuators. (Credited to the author.)

seal. During the seal test, a pressure reduction of 3% of the test pressure may be allowed. Therefore, for an actuator seal test at 32 bar (0.2*160 = 32 bar) for three minutes, the maximum allowable pressure reduction is $0.03 \times 32 \times 3/60 = 0.048$ bar.

9.6.2 Electrical Actuators Testing

Electrical actuators do not utilize pressurized fluid, so they do not require pressure tests. Electrical actuators are subjected to standard testing procedures which include operating them locally and remotely as well as measuring the force or torque produced by them.

QUESTIONS & ANSWERS

1. Which of the following statements are correct about testing pipeline valves?

   A. Production testing of pipeline valves is conducted after the valves are manufactured and assembled.
   B. In the event that a valve fails during testing, it must be replaced with a new and identical valve.
   C. Testing of pipeline valves can reduce OPEX or operation costs.
   D. All tests for pipeline valves are performed at extremely low temperatures.

   Answer) Option A is the correct choice. It is incorrect to choose option B because if a pipeline valve fails during testing, the problem should be detected and rectified before the valve is tested again. Option C is correct, but all of the tests for pipeline valves are conducted at room temperature, so option D is incorrect. In conclusion, options A and C are the correct answers.

2. Choose the correct sentence regarding testing pipeline valves.

   A. For pneumatic and hydrostatic closure tests, the test pressure is equal.
   B. In accordance with API 6D, the minimum hydrostatic seat test duration is longer than the minimum pneumatic or gas seat test duration.
   C. Leak rate A according to ISO 5208 is usually not applicable to pipeline valves.
   D. It takes 20 minutes to perform a hydrostatic shell test on a 30" pipeline valve.

   Answer) Option A is incorrect since the test pressure for pneumatic closure tests is typically between 4 and 7 bar, also known as low pressure. On the other hand, the hydrostatic closure test value is 1.5 times the nominal pressure of the valve, which is referred to as high pressure. According to API 6D and Table 9.4, option B is incorrect too, because the minimum duration of a pneumatic or gas seat test is longer than that of a hydrostatic test. The hydrostatic seat or closure test duration for a 30" pipeline valve is 10 minutes. Alternatively, the pneumatic or gas seat test lasts 30 minutes, which is three times longer than the hydrostatic seat test. The correct answer is C because pipeline valves have metallic seats subject to leakage; however, leak rate A indicates there has been no leakage. Option D is incorrect, and the hydrostatic shell test duration for a pipeline valve of 30" is 30 minutes.

3. What are the reasons behind the rejection of the 20" CL1500 pipeline valve with A105 body material and DIB2 configuration during testing?

   A. Although the valve datasheet specifies a maximum opening torque of 73,344 Nm, the opening torque measured during the test was 90,000 Nm.
   B. When the cavity pressure reaches 285 bar, the self-relief seat of the valve releases the pressure.
   C. Hydrostatic pressure is significantly reduced during the shell test, and an inspector notices some moisture on the valve's external surface.
   D. Leakage from the DPE seat during the cavity test is measured as leak rate B, whereas the allowable leakage from the DPE seat during the hydrostatic closure test is rate C.

   Answer) According to Option A, the valve should be rejected due to the results of the function test. The measured maximum opening torque exceeds the valve datasheet's maximum opening torque. Option B does not indicate a specific condition for valve rejection. As shown in Table 9.2, the valve body is made of A105 with a nominal pressure of 255 bar. The SR seat should release the cavity pressure at a value less than 339.15 bar (1.33*255 = 339.15), and it operates at 285 bar. Option C indicates that the valve should be rejected. Option D does not result in valve rejection because a double piston effect (DPE) seat has less leakage from the cavity than allowed leakage from the DPE seat during a hydrostatic cavity test. Therefore, options A and C are correct.

4. Which test parameter is different between hydrostatic and pneumatic closure tests?

   A. Type of test fluid
   B. Test pressure
   C. Leakage rate
   D. All options are correct

   Answer) For the hydrostatic closure test, the test fluid is water, while for the pneumatic closure test, the test fluid is air or gas, so option A is correct. For the hydrostatic closure test, the test pressure is high and equal to 1.1 times the valve's nominal pressure. In contrast, the test pressure of a pneumatic closure is low and lies within a range of 4–7 bar. Therefore, option B is also acceptable. It is correct to select option C since the allowable leakage rates for the closure test differ between liquids and gases as shown in Table 9.5. Since all three choices are correct, option D should be selected.

5. Fill in the blanks in the following paragraph about pipeline valves testing.
   It is ________ for the valve manufacturer to test all valves internally in advance before the attendance of the external inspector. The hydrostatic shell test is performed when the valve is in a ________ position. The seat test also called a ________ is conducted to examine the sealing capacity of the valve between the closure member and seats. A function test includes both ________ and ________. A backseat test is only applicable to ________ pipeline valves. Applying a ________ is very hazardous, and personnel involved, such as operators and

inspectors, shall take appropriate safety measures during this test. Test pressure for a high-pressure gas test applied on a valve with a nominal pressure of 259 bar is _______ bar.

A. Optional, fully open, drift test, pressure test, torque measurement, ball, low-pressure air test, 290
B. Mandatory, half-open, leakage test, torque measurement, operating time measurement, TCG, high-pressure gas test, 285
C. Optional, fully closed, closure test, operating time measurement, pressure test, TCG, hydrostatic test, 300
D. Mandatory, half-open, closure test, pressure and temperature tests, ball, high-pressure gas test, 260

Answer) Option B is the correct answer. It is mandatory for the valve manufacturer to test all valves internally in advance before the attendance of the external inspector. The hydrostatic shell test is performed when the valve is in a half-open position. The seat test, also called a leakage or closure test, is conducted to examine the sealing capacity of the valve between the closure member and seats. A function test includes both torque and operating time measurements. A back-seat test is only applicable to TCG pipeline valves. Applying a high-pressure gas test is very hazardous, and personnel involved, such as operators and inspectors, shall take appropriate safety measures during this test. Test pressure for a high-pressure gas test applied on a valve with a nominal pressure of 259 bar is 285 bar.

Note: According to API 6D, the test pressure for a high-pressure gas test, whether on the body or seats of a pipeline valve, is equal to 1.1 times the nominal pressure (1.1*259 = 285 bar).

6. In order to ensure the safety and reliability of valves during a fire, what types of tests are conducted?

A. Fire test
B. Antistatic test
C. Electrical continuity test
D. All options are correct.

Answer) All tests are applicable, so option D is the best choice.

7. Three pipeline valves made of ASTM A105 bodies are tested during the pressure test. The first valve has a size of 20" and is of pressure class 1500, the second valve has a size of 30" and is of pressure class 2500, and the third valve has a size of 38" and is of pressure class 1500. The cavity relief pressures during the cavity test for these three valves are 361 bar, 560 bar, and 320 bar, respectively. Which test results are acceptable? (Assume that the design pressure of the valves is equal to the PN)

Answer) According to Table 9.2, the first valve has a PN of 255 bar, and the design pressure is 255 bar. The cavity release pressure is 361 bar > 1.33 × 255 bar (design pressure) = 339.15 bar; therefore, the test result for the first valve is unacceptable.

The second valve is of class 2500 with PN 425 bar, which corresponds to a design pressure of 425 bar. It is calculated that the cavity release pressure is 560 bar < 1.33 × 425 bar (design pressure) = 565.25 bar. Accordingly, the test result for the second valve is acceptable.

The third valve is a class 1500 valve with a PN of 255 bar, and its design pressure is 255 bar. Cavity release pressure is 320 bar < 1.33 × 255 bar (design pressure) = 339.15 bar. Thus, the test result for the third valve is satisfactory.

Therefore, the cavity test results for the second and third valves are acceptable.

8. A 40" CL1500 pipeline ball valve is subjected to three types of closure tests: hydrostatic and low-pressure air tests with maximum permissible leakage rates B, and high-pressure gas tests with maximum permissible leakage rates C. What are the maximum permitted leakage rates in cubic millimeters during each test?

Answer) The 40" valve is equal to the nominal diameter (DN) 1000. In accordance with ISO 5208 leak rate B, the maximum allowable leakage rate for water and air is equal to 0.01*DN (0.01*1000 = 10 $\frac{mm^3}{second}$) and 0.3*DN (0.3*1000 = 300 $\frac{mm^3}{second}$), respectively. In accordance with Table 9.4, the duration of the hydrostatic and air seat/closure tests is 10 minutes and 30 minutes, respectively. For these two tests, the following formula can be used to determine the maximum allowable leakage during the test period:

The maximum leakage of the 40" valve during a hydrostatic closure test lasting 10 minutes = $10 \frac{mm^3}{second} * 60 \frac{second}{minute} * 10 \text{ minutes} = 6000 \text{ mm}^3$

The maximum leakage for the 40" valve during a 30-minute low-pressure air closure test = $300 \frac{mm^3}{second} * 60 \frac{second}{minute} * 30 \text{ minutes} = 540{,}000 \text{ mm}^3$

According to ISO 5208, the maximum leakage rate C for high-pressure gas is equal to 3*DN (3*1000 = 3000 $\frac{mm^3}{second}$)

The maximum amount of leakage during the closure test of the 40" valve under high pressure for a period of 30 minutes = $3000 \frac{mm^3}{second} * 60 \frac{second}{minute} * 30 \text{ minutes} = 5{,}400{,}000 \text{ mm}^3$

BIBLIOGRAPHY

1. American Petroleum institute (API) 598. (2004). *Valve inspection and testing*, (18th edition). Washington, DC: ASME.
2. American Petroleum Institute (API) 6FA. (2011). *Specification for fire test for valves*, (3rd edition). Washington, DC: ASME.
3. American Petroleum institute (API) 6D. (2014). *Specifications for pipeline and piping valves*, (24th edition). Washington, DC: ASME.

4. American Petroleum Institute (API) 607. (2016). *Fire test for quarter-turn valves and valves equipped with non-metallic seat*, (7th edition). Washington, DC: ASME.
5. American Society of Mechanical Engineers (ASME) B16.34. (2017). *Valves–Flanged, threaded, and welding end.* New York, NY: ASME.
6. International Organization of Standardization (ISO) 14313. (2007). Petroleum and natural gas industries – Pipeline transportation systems– Pipeline valves, (2nd edition). Geneva, Switzerland.
7. International Organization of Standardization (ISO) 10497 (2010). *Testing of valves – Fire type testing requirements*, (3rd edition). Geneva, Switzerland.
8. Sotoodeh K. (2019). Valve operability during a fire. *American Society of Mechanical Engineers (ASME), Journal of Offshore Mechanics and Arctic Engineering*, 141(4), 044001. https://doi.org/10.1115/1.4042073. Paper No. OMAE-18-1093.
9. Sotoodeh K. (2020). Subsea valves and actuators: A review of factory acceptance testing (FAT) and recommended improvements to achieve higher reliability. *Journal of Life Cycle Reliability and Safety Engineering, Springer*. https://doi.org/10.1007/s41872-020-00153-w
10. Sotoodeh K. (2022). *Cryogenic valves for liquified natural gas plants*, (1st edition). USA: Elsevier (Gulf Professional Publishing).

10 Preservation and Packing

10.1 INTRODUCTION

In order to protect them from harmful environmental conditions, valves, including pipeline valves, must be preserved throughout and after the delivery process. Examples of harmful environmental conditions include wind, dust, humidity, salt, sandblasting and painting, high and low environmental temperatures, and external forces. Protection of valves includes transportation, storage, handling, installation, and testing in the construction yard. Depending on the storage and fabrication time, the preservation period can range from six months to two or even five years. According to project specifications, it is the valve manufacturer's responsibility to execute initial valve preservation. Valve suppliers or manufacturers should provide purchasers with all the necessary documentation, including preservation procedures. These procedures will describe how valves should be stored and installed in accordance with preservation requirements. The valve supplier/manufacturer should also prepare datasheets for all preservative chemicals and send them to the purchaser for approval. After the valve pressure test and during the final inspection, an inspector checks the majority of preservation requirements.

10.2 PRESERVATION TYPES

This chapter divides preservation activities into three categories: general, internal, and external.

10.2.1 General Preservation

To prevent corrosion, it is necessary to limit the chloride content in the test water to 30 part per million (ppm). In addition, add a corrosion inhibitor to the hydrotest water, drain the test water from the valve, and dry the valve after the test. For carbon steel valves such as pipeline valves with lower corrosion resistance, this point is essential. A case study is presented in this section, in which test water is not drained from three pipeline valves after factory acceptance tests (FATs) and causes corrosion inside a couple of the valves. This review starts with the emergency shutdown (ESD) valve installed as the last or third valve on the offshore pipeline. The drain flange installed at the bottom of the valve was opened by personnel on site (see Figure 10.1). In the picture, the drained water was mixed with a dark and brown substance that looked like grease and oil. Figure 10.2 shows that greasier product has been deposited on the drain flange face of the valve's body. Whether the deposited product was grease, such as preservatives used inside the valve, or corrosion product, or a mixture of both, is the main question. Personnel cleaned the flange face using tissue and solvent. Solvent and tissue cleaning removed some dark and brown products, showing that the extracted products are only dirt and grease. Some products were

DOI: 10.1201/9781003343318-10

FIGURE 10.1 Drain flange opening at bottom of the first pipeline valve. (Photo by the author.)

FIGURE 10.2 Greasy product was deposited on the drain flange face of the ESD valve's body. (Photo by the author.)

not properly cleaned with solvent and tissue, showing signs of corrosion and rust. Figure 10.3 illustrates the drain hole at the bottom of the valve connected to the cavity blind flange. Furthermore, it was covered with a mixture of greasy preservatives and corrosion products. In this case study, SOCOPAC was used to protect the valve from corrosion. This was done by forming an effective barrier against the agents responsible for corrosion such as water, oxygen, acids, and salts. SOCOPAC is resistant to heavy rains, corrosive offshore environments, and tropical weather. The film of this preservative is medium-hard, wax-like, and dry to the touch. According to this study, two conclusions can be drawn. The first is that draining the water from the valve after hydrostatic pressure testing could cause corrosion even though the

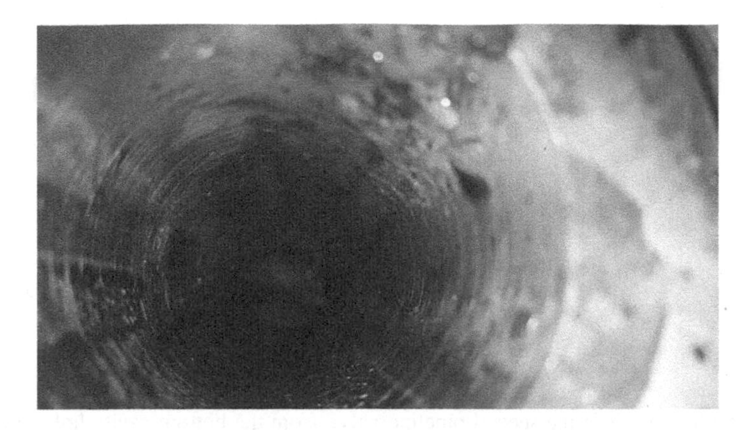

FIGURE 10.3 The drain hole at the bottom of the ESD valve is connected to the drain blind flange. (Photo by the author.)

FIGURE 10.4 The blind flange of the second pipeline valve with a sample of the greasy deposits. (Photo by the author.)

chloride content of the test water was limited to 30 parts per million (ppm) and corrosion inhibitors were added. In addition, SOCOPAC was not effective enough to prevent corrosion as a strong preservative.

The second pipeline valve from which the test water was drained after the test was inspected by an inspector. Figure 10.4 illustrates how the blind flange was removed from the bottom of the cavity drain. As can be seen in the photograph, samples of grease had been collected, which had been used during assembly to place the ring gasket inside the grove of the flange. A tissue and solvent were used to remove the grease, and the valve was inspected afterward. Figure 10.5 illustrates the inside of the cavity drain hole for the second pipeline valve. The valve was generally clean, with only traces of grease and no corrosion products.

FIGURE 10.5 Inside of the second pipeline valve from the bottom cavity hole. (Photo by the author.)

FIGURE 10.6 Limit switch box mounted on a hydraulic actuator. (Photo by the author.)

10.2.2 Internal Preservation

1. Any foreign materials should be removed from the internal surface of the valves.
2. Electrical components of actuators, such as the limit switch box, solenoid valves on the control panel, and junction boxes must be preserved with VCI or VpCI (vapor corrosion inhibitor, also known as volatile corrosion inhibitor) or any other similar preservative compounds to prevent rust and corrosion. The vapors released by VCIs provide corrosion protection by forming a thin layer on the metal surface. In hydraulic actuators, a limit switch box (see Figure 10.6) contains a limit switch that controls and limits the position of the valve closure member (e.g., ball) at open and closed positions. Suppose the pipeline valve is completely open when the ball's hole is parallel to the fluid flow passing through the valve. A valve is fully closed when the ball's hole forms a 90° angle with the flow direction. At fully open and

fully closed positions, the limit switch controls the ball's stoppage. During hydraulic actuator operation, solenoid valves controlled by electrical signals are installed inside the control panels of hydraulic actuators (see Figure 10.7) to permit and close hydraulic fluid flow. An electrical actuator's junction box contains the electrical connections, as shown in Figure 10.8.

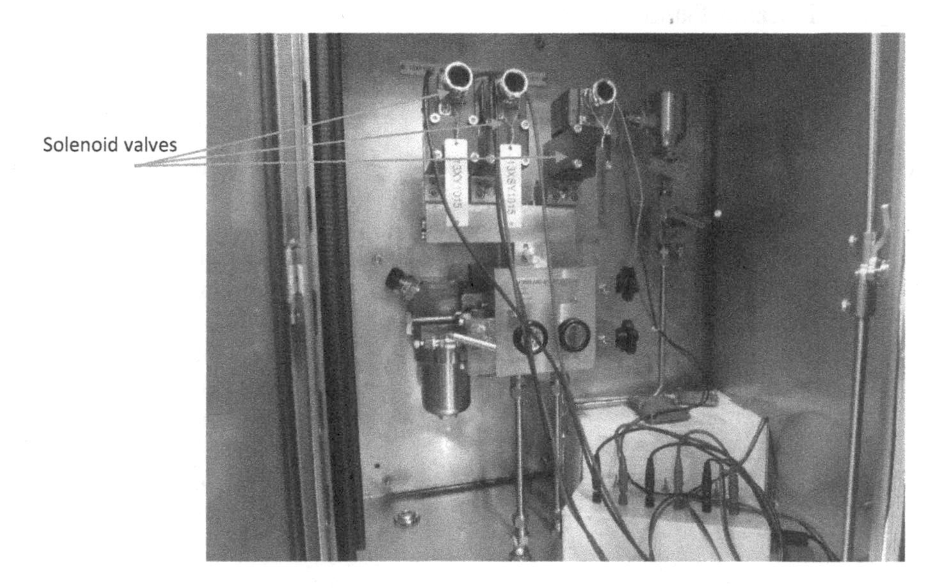

FIGURE 10.7 Typical solenoid valves inside the control panel of a hydraulic actuator. (Photo by the author.)

FIGURE 10.8 An electrical actuator's junction box. (Photo by the author.)

3. The author recommends corrosion inhibitors such as Tectyl, VCI, VpCI, and SOCOPAC for internal corrosion protection of pipeline valves. Tectyl is a preservative used to protect metals against rust and corrosion, moisture, and chloride attack.

10.2.3 External Preservation

1. In Chapter 2, it was discussed how pipeline valves' bodies and pup pieces are coated to prevent corrosion in offshore environments. Nevertheless, the end of the pup piece that has to be welded to the pipeline in the construction yard needs to remain uncoated since welding can damage or weaken the coating. An uncoated portion of the pup piece can be protected from external corrosion during transportation and handling before welding to the pipeline by a layer of primer. An alternative method is the application of a preservative oil, such as VpCI, to the uncoated surfaces of pup pieces at the valve factory.
2. Pup pieces' bevel edges should be protected by plywood rounded sheets of 10 mm or plastic caps during transportation. A plywood sheet can damage the end of the pup piece, so a ring made of nitrile rubber (NBR) is placed between the plywood and pup piece end to prevent possible damage. Around the plywood, NBR, and the bevel end of the valve's pup piece, an oil-resistant rubber tape is applied. As shown in Figure 10.9, an ESD valve

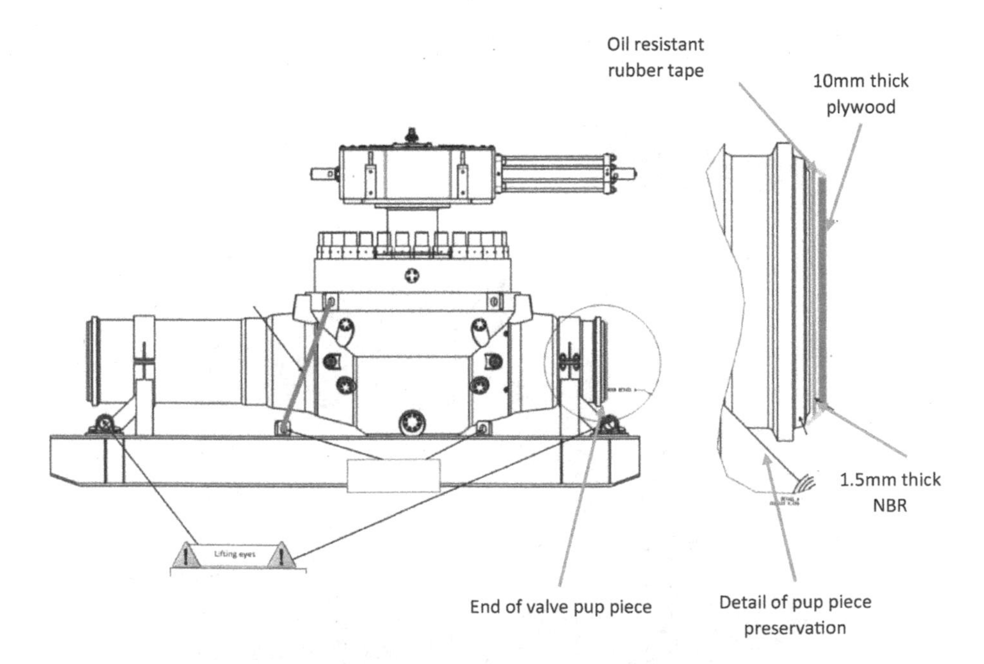

FIGURE 10.9 Detail of the pup piece end protection belonging to a hydraulically operated pipeline valve. (Photo by the author.)

with a hydraulic actuator is preserved using the preservation techniques described above and is placed on a skid for transportation. The pup piece bevel end protection detail is shown on the right side of the picture by a plywood sheet and an NBR.

An illustration of pipeline valves on a truck with electrical actuators is shown in Figure 10.10. The valve ends are blinded with round plywood sheets attached by wrapped oil-resistant tapes to the valve pup pieces.

3. Glass components such as screens attached to electrical actuators (see Figure 10.11) should be protected with plywood.
4. A preservative oil should be sprayed onto any external threads on the valve or actuator and taped to prevent corrosion.

FIGURE 10.10 Transporting two electrically actuated pipeline valves on a truck with plywood rounded sheets wrapped with tape to conceal the ends of the pup pieces. (Photo by the author.)

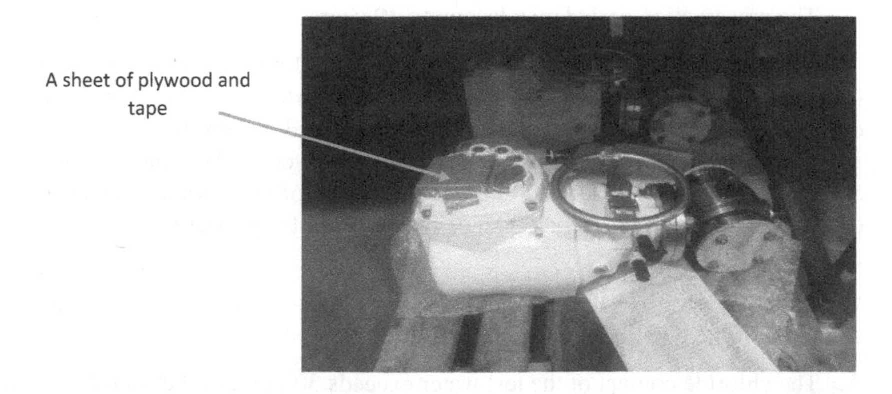

FIGURE 10.11 Protecting an electrical actuator screen with plywood. (Photo by the author.)

FIGURE 10.12 Placing a wooden frame under a pipeline valve during transportation on a truck. (Photo by the author.)

10.3 PACKING

Valve packing should be appropriate to ensure safe transportation to the customer's location and should be based on the customer's order and requirements. Industrial valves are frequently packed in wooden boxes. However, as shown in Figure 10.10, pipeline valves with the associated actuators may be placed on and fastened to wooden or metal skids during transportation. It is not common to transport such large pipeline valves in wooden boxes. In most cases, pipeline ball valves are transported with the ball in the open position. An appropriate connection between the valve and actuator should be robust enough to prevent the valve and actuator from being disassembled. For transportation, the wooden frame shown in Figure 10.12 is located under the top-entry ball valves.

QUESTIONS & ANSWERS

1. Find the correct sentence regarding the preservation of pipeline valves.

   A. Pipeline valves are only preserved when the valves are in the factory.
   B. Valve preservation is limited to internal protection.
   C. Manufacturers of pipeline valves prepare the preservation procedure in accordance with the project specifications.
   D. The preservation period may last up to 30 years.

   Answer) Option A is incorrect since all valves, including those installed on pipelines, should be protected during all stages of transportation, storage, lifting, handling, installation, and testing in the construction yard. Furthermore, option B is also incorrect since preservation of pipeline valves could be general, internal, or external, and it is not limited to the internal type of preservation. The correct option is C. Option D is incorrect since preservation for 30 years is too long and costly.

2. What are the conditions under which the hydrostatic test water can cause internal corrosion of the pipeline valve being tested?

   A. The chloride content of the test water exceeds 30 ppm, or the limit provided in the engineering documents.

B. The test water is not treated with corrosion inhibitors.
C. The test water is not drained from the valve following the test.
D. All choices are correct.

Answer) Since all three conditions can cause internal corrosion of the pipeline valve, option D is the correct answer.

3. Select the external preservation option for pipeline valves.

A. Using vapor corrosion inhibitor (VCI) for the electrical actuators' junction boxes
B. Protecting the valve weld ends with plywood
C. Drying the valve from the water after performing a pressure test
D. Adding corrosion inhibitor to the test water

Answer) Option A is an internal preservation strategy, so it is incorrect. The correct answer is B. Both options C and D are considered general preservation methods, so they are both incorrect.

4. What are the functions of plywood sheets in the preservation of valves?

A. Avoiding the ingress of foreign objects or materials
B. Providing easy connection between the valve and connected piping
C. Protecting the valve's bevel ends
D. Lubrication of the valve

Answer) Both options A and C are correct.

5. Which electrical component should be protected from external corrosion by VCI?

A. Hydraulic actuators' solenoid valves
B. Junction boxes for electrical actuators
C. Limit switches for hydraulic actuators
D. All options are correct

Answer) Option D is correct.

BIBLIOGRAPHY

1. Sotoodeh K. (2016). Valve preservation requirements. *Valve World Magazine*, 21(05), 53–56.
2. Sotoodeh K. (2021). *A practical guide to piping and valves for the oil and gas industry*, (1st edition). Austin, USA: Elsevier (Gulf Professional Publishing).

11 Handling, Lifting, and Transportation

11.1 HANDLING

In this chapter, handling a valve means preparing the valve for lifting, movement, and transportation. Industrial valve handling methods are divided into two categories: the packed valves inside crates, cases, or wooden boxes, and the second is related to unpacked valves such as those installed on pipelines.

With the aid of forklift trucks equipped with proper fork hitches, valves packed in crates or cases are lifted and handled. The valves inside crates and cases are equipped with lifting lugs that can be used for lifting and transportation by considering the center of gravity (COG) mark on the valve. The COG is the single point of an object, such as a valve, where all the weight of the object is concentrated. Why is it so vital to consider the COG? A COG location is crucial to the stability of an object. When a crane or other lifting machine lifts an object, such as a valve, the valve is free to tilt and move in most simple lifting ways until the COG of the valve or other object is directly under the crane hook. It means that if the COG of the object or valve can be calculated or known, it is possible to lift the valve in a way that prevents tilting during the lifting and handling process. Handling and lifting packed valves must be done safely and according to safety regulations.

The following paragraph summarizes the requirements for handling and transporting pipeline valves that are typically unpacked. Pallets must be used to lift and handle unpacked valves, and the machined surfaces of the valves must be protected to prevent damage from external sources. Pipeline valves have enormous dimensions, so tools such as hooks, fasteners, chains, and ropes are used to handle these valves, prevent them from falling, and handle them during the lifting, handling, and transportation. The chains, ropes, and cables used to lift and handle the valves must be strong enough and sized correctly so that they can be handled safely and appropriately. The first point to remember regarding lifting pipeline valves is that the valve should not be lifted using only one lifting lug, as pipeline valves generally contain four lifting points. Second, the actuator lifting lug should only be used to lift the actuator only, and not the assembled valve and actuator. As the actuator manufacturer designs the actuator lifting lugs to be used only for lifting the actuator, they are not strong enough to handle the combined weight of the valve and connected actuator. Pipeline valves are best handled using slings attached to four lifting lugs of the valve to lift the valve vertically as shown in Figure 11.1. In order to prevent any rotation of the valve assembly, check carefully that the actuator installed on the top of the valve is entrapped between the slings.

DOI: 10.1201/9781003343318-11

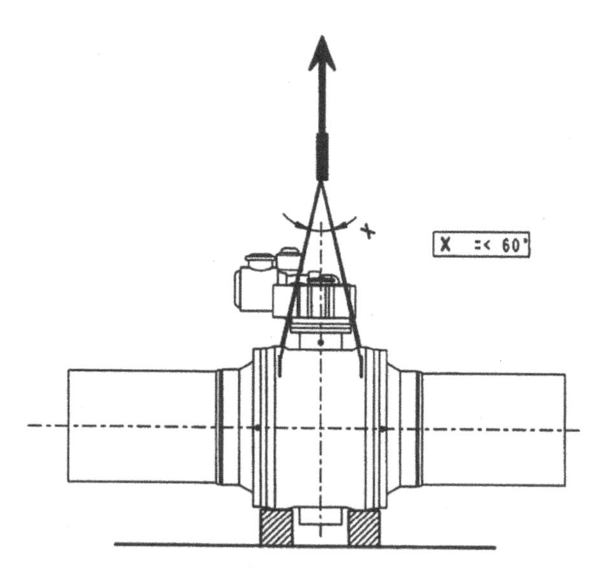

FIGURE 11.1 Handling a pipeline valve by attaching slings to four lifting lugs. (Photo by the author.)

11.2 LIFTING

Each pipeline valve is lifted by means of four lifting lugs, two of which are highlighted in Figure 11.2 and two others are located on the valve's backside, not shown in the picture.

As shown in the picture, the valve is fastened to a skid during lifting and transportation. This is explained in additional detail in Section 11.3. All valve lifting lugs must be certified by the valve manufacturer to ensure they are able to lift and handle the total weight of the assembled valve and actuator. The pipeline valve is raised

FIGURE 11.2 A hydraulically actuated pipeline valve, complete with lifting lugs that provide easy lifting and transportation. (Photo by the author.)

FIGURE 11.3 Lifting the pipeline valve with a crane utilizing chains and an H-shaped device. (Photo by the author.)

using a yellow H-shaped device, also known as an H-frame, and this tool is considered a special lifting tool. In the absence of an H-shaped device, the chains could collide with the actuator on top of the valve during lifting. It is not recommended to dismantle the actuator from the valve during transportation. As a result, the actuator has to be reassembled on the valve in the construction yard, and the actuator and valve have to be tested together, which adds time and cost to the project. Figure 11.3 shows a crane used to lift pipeline valves using chains and an H-shaped device. During lifting, the H-shaped tool is attached to the lifting lugs of four valves by four chains. In this case, the H-shaped frame weighs 8 tons, and it has a maximum load capacity of 100 tons. Two adjustable two-leg slings are connected to the H-shaped frame from the top in an effective length of 2.7 meters, and four adjustable single-leg slings are connected to the frame from the bottom in an effective length of 5 meters.

The following are some general lifting requirements and considerations:

- Ensure that the operators of lifting equipment are skilled enough to perform their duties.
- Do not overload the lifting equipment with more than it can handle.
- Do not work, stand, or pass under the suspended pipeline valves.
- Do not use the lifting equipment while talking or texting on your mobile phone.
- Be aware of any electrical hazards in the lifting area.
- Take all necessary precautions to avoid risks when lifting.
- In the event of an unsafe situation, stop working.
- Ensure the crane is properly inspected before operation.
- In addition to the crane, the operator or a qualified inspector should visually inspect the hoist wire, rope, chain, and other hardware.

11.3 TRANSPORTATION

The purpose of this section is to define and describe the method and safety procedures followed by a transportation company in transporting three huge pipeline ball valves for an offshore Norwegian project from the manufacturer's factory in France to Korea, where installation and construction took place. In order to achieve a successful transportation operation, it is crucial to use suitable transportation tools and facilities that comply with national and international regulations, practices, safety methods, etc. Therefore, the proposed transport, lifting, and lashing methods are applied here to three pipeline ball valves; one hydraulically actuated valve weighing 79 tons (see Figure 11.4) as well as two identical electrically actuated valves weighing almost 70 tons (see Figure 11.5). Lashing refers to the attachment of chains, wires, and ropes with linking devices utilized during handling, lifting, and transportation to secure cargo (e.g., valves and actuators). In Figure 11.5, four ratchet load binders are used to connect the valve lifting lugs to the lifting points on the skid. This keeps the valve firmly attached to the skid during handling and lifting. In addition, there are two collars at both ends of the skid to secure the pup piece. There is a rubber seal at the bottom of the valve to prevent friction with the skid. Ratchet load binders combine a ratchet handle with two tension hooks at either end. Hooks are attached to chains while the load binder tightens the chains.

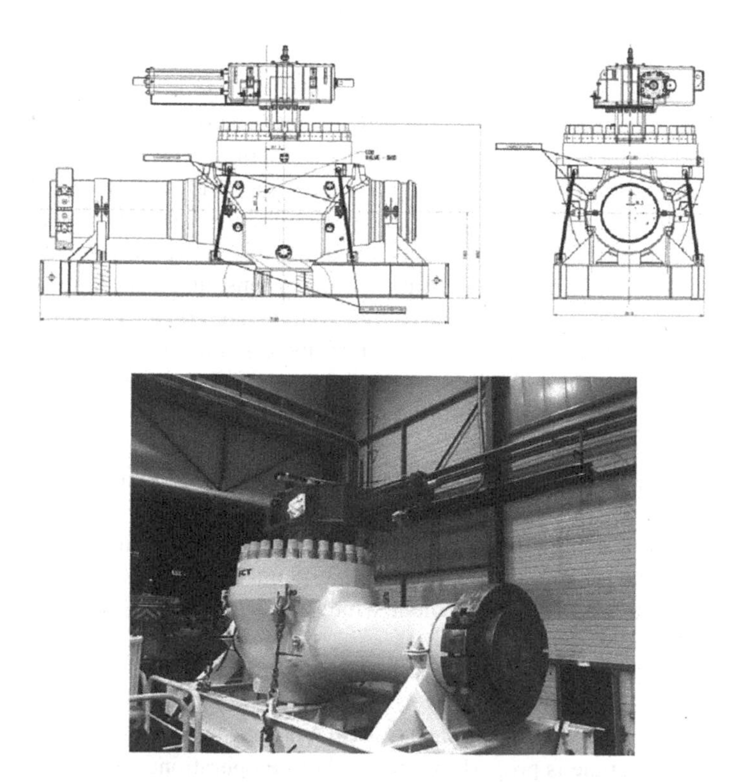

FIGURE 11.4 A hydraulically actuated pipeline ball valve is ready for transportation (the total weight of the valve and actuator = 79 tons). (Photo by the author.)

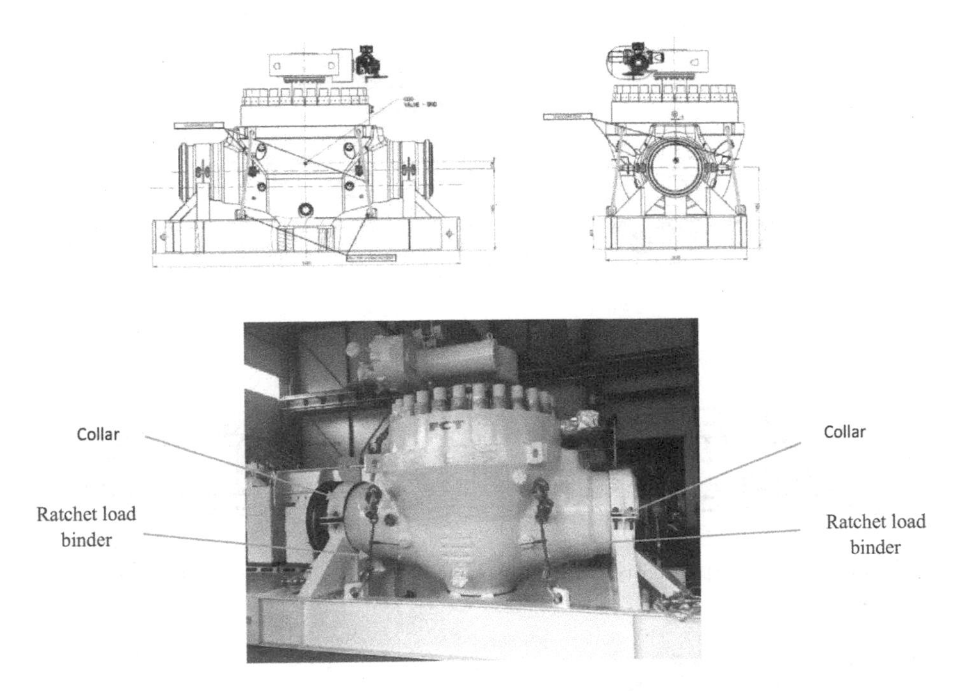

FIGURE 11.5 A pipeline valve with electrically actuated actuators, ready for transportation (weight of the valve and actuator: approximately 70 tons). (Photo by the author.)

The valves are transported by a combination of land, road, and marine methods. In Figure 11.6, we show the transportation of the hydraulically operated pipeline ball valve by truck. In Figures 11.7 and 11.8, four lashing chains are used to secure the valve firmly on the skid or back of the truck during transportation.

In Figures 11.7 and 11.8, each lashing chain has a nominal size of 13 mm and a lashing capacity of 20 tons. Consequently, four lashing chains with a lashing capacity of 80 tons can be used to secure a hydraulically operated actuator weighing 79 tons. An important component of transportation is a D-ring, which is a part of the skid used to secure the valve and lift the skid. As shown in Figure 11.7, the skid is equipped with four ***D-rings***; each has a load capacity of 36 tons, and a length of 185 millimeters and a thickness of 5 millimeters of the ring at the bottom are welded to the skid (see Figure 11.9 for details concerning the D-ring). Figure 11.10 illustrates the pipeline ball valve installed on the skid; the skid is equipped with eight welded stoppers with faceplates that are used to prevent the skid from moving during transportation. There are two plates in the stopper arrangement: a face plate and a stopper plate that are perpendicular and welded together. The face plate is parallel to the skid, whereas the stopper plate prevents lateral movement (see Figure 11.11 for details on the stopper).

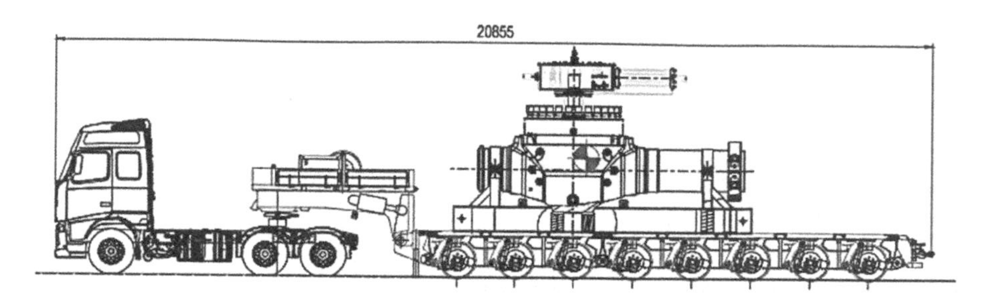

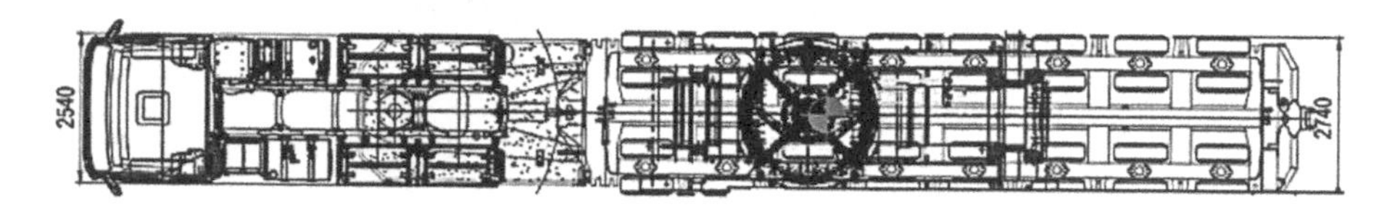

FIGURE 11.6 Transport of hydraulically operated pipeline ball valves by truck. (Photo by the author.)

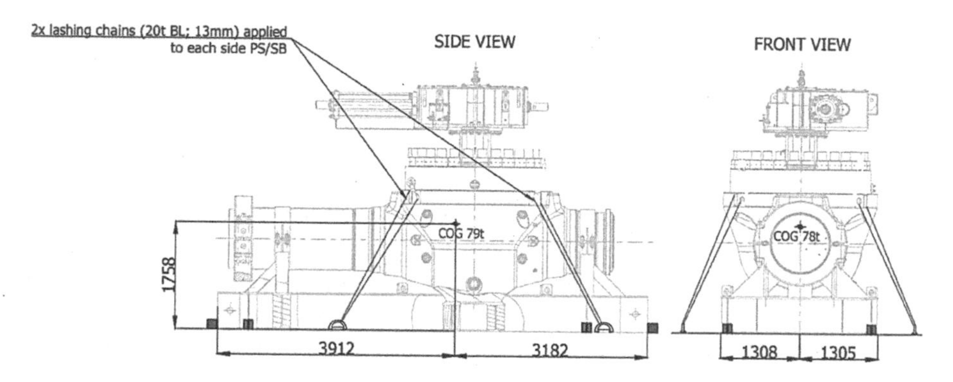

FIGURE 11.7 Four lashing chains were used to attach the hydraulically operated pipeline valve to the skid. (Photo by the author.)

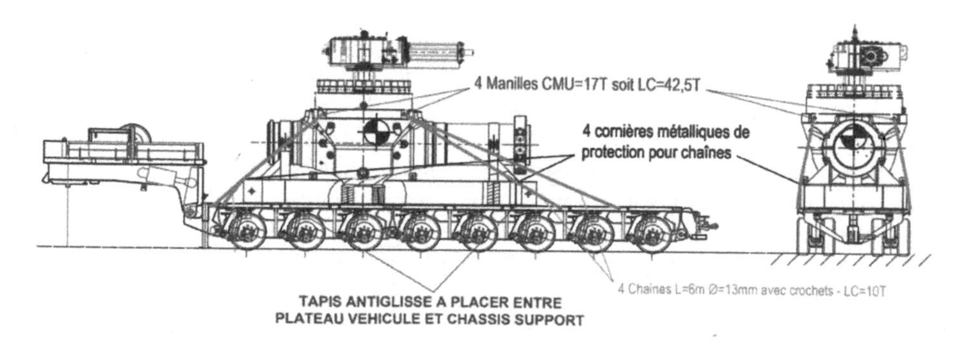

FIGURE 11.8 Connecting the hydraulically operated pipeline valve to the truck using four lashing chains. (Photo by the author.)

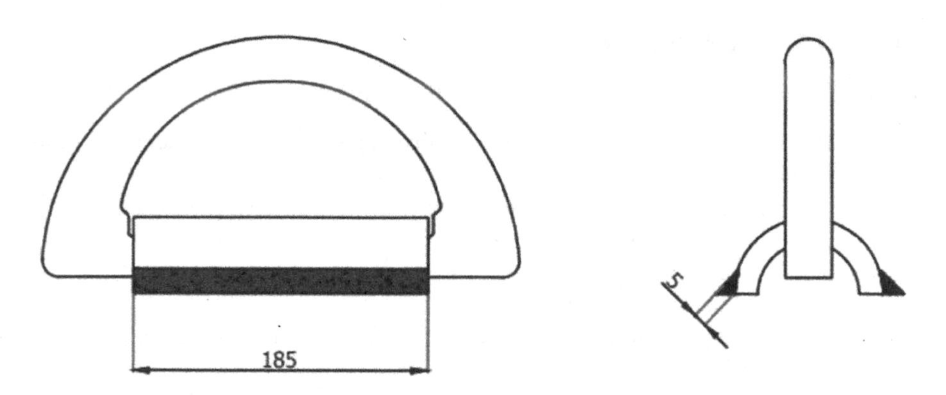

FIGURE 11.9 A D-ring detail. (Photo by the author.)

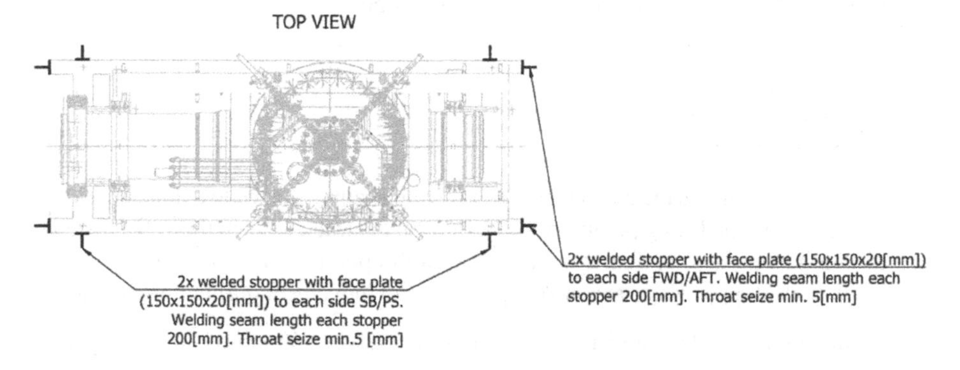

FIGURE 11.10 Top view of a pipeline valve on the skid with eight stoppers. (Photo by the author.)

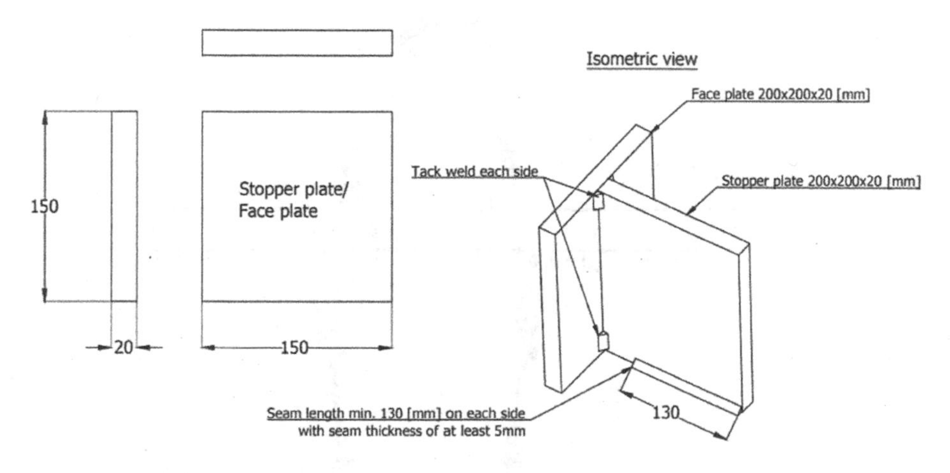

FIGURE 11.11 Stopper arrangement for a skid, including the face and stopper plates. (Photo by the author.)

QUESTIONS & ANSWERS

1. Which sentence is not correct regarding the handling of pipeline valves?

   A. Pipeline valve handling consists of preparing them for movement and transportation.
   B. A pipeline valve is considered an unpacked valve during handling, lifting, and transportation.
   C. The lifting personnel should not use an actuator lifting lug to lift the total weight of the valve and actuator.
   D. By attaching a lashing chain to only one lifting lug, the lifting personnel can lift the valve.

   Answer) Since lifting personnel must use all four lifting lugs or points to handle, lift, and transport pipeline valves, option D is incorrect.

2. Lifting the skid is accomplished by which component?

   A. D-ring
   B. Lashing chains
   C. Stopper face plate
   D. Stopper plate

   Answer) Option A is the correct answer. Lashing chains are used to connect the valve lifting lugs to the lifting points on the skid or on the truck to keep the valve fastened during transport, so option B is incorrect. Both options C and D are incorrect because both stopper plates are used to prevent skid movement during transportation.

3. Figure 11.12 shows a tool for lifting pipeline valves. Which of the following is not correct about it?

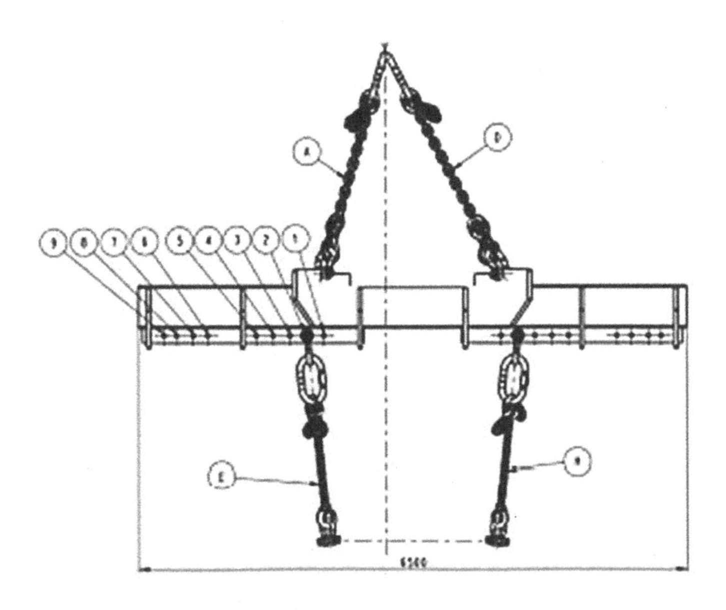

FIGURE 11.12 A special tool. (Credited to the author.)

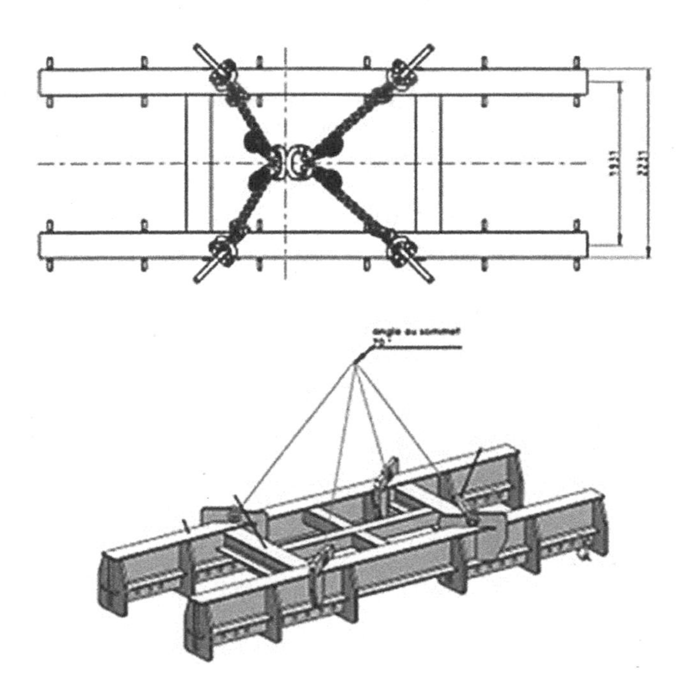

FIGURE 11.12 (Continued)

A. The special tool is known as the H-shaped tool or H-shaped frame.
B. The tool is mainly used during the handling and transportation of pipeline valves.
C. A tool with a load capacity of 100 tons can lift a pipeline valve weighing 70 tons.
D. In the picture, the tool is attached to the valve by four adjustable slings.

Answer) The wrong answer is B because the special tool called the H-shaped tool or frame is primarily used to lift pipeline valves and not to handle or transport them. Options A, C, and D are all correct answers.

4. What component secures a pipeline valve pup piece to a skid?

A. Skid lifting lugs
B. Lashing chains
C. Collars
D. Ratchet load binder

Answer) The correct answer is C. Skid lifting lugs are used to lift the skid or to fasten the valve to the skid with chains and ratchet load binders, so option A is incorrect. Lashing chains and ratchet load binders are both used to secure the valve to the skid or to the truck, so options B and D are incorrect.

BIBLIOGRAPHY

1. Pbctoday. (2019). How centre of gravity affects a lift (in simple terms). [online] https://www.pbctoday.co.uk/news/industry-insight/centre-gravity-affects/57157/ [access date: 15th January 2022].
2. Sotoodeh K. (2021). A practical guide to piping and valves for the oil and gas industry, (1st edition). Austin, USA: Elsevier (Gulf Professional Publishing).

12 Installation, Operation, and Maintenance

12.1 INSTALLATION AND OPERATION

12.1.1 Storage Before Installation

In some cases, valves must be stored prior to installation while taking into consideration the following essential requirements:

- The valve must be stored in a clean, dry, and enclosed space.
- During preservation, plastic caps or plywood end protection sheets must remain attached to the valves. All original valve protections should be maintained during storage.
- In order to ensure that the storage area remains clean and dry, onsite personnel should conduct periodic inspections and checks. In addition, the onsite personnel are responsible for checking the packaging of valves and actuators.
- Storage in an open area prior to installation is permitted for a limited period of time when the valve is appropriately packed. Even for a short time, open area storage is not permitted for pipeline valves since they are not packed in wooden boxes.

12.1.2 Valve Installation and Operation on the Pipeline

A valve inspection and some actions should be performed prior to valve installation on the pipeline as follows:

- The onsite personnel responsible for installing pipeline valves shall remove the valves from the shipping skid or frame properly to avoid damaging the valves and actuators.
- The end plywood sheets fastened in place with the wrapped tapes for packing and preservation and explained in more detail in Chapter 10 shall be removed before the valve installation.
- The inside of the valves shall be cleaned to ensure that there are no solid objects or particles inside that can damage the soft (non-metallic) seals located between the seats and body of the valves.
- The valve bevel (welded) ends shall be inspected to ensure no damage on them before being welded to the pipeline. In addition, the pipeline ends to be welded to the valves shall be checked to make sure that there is no damage.
- The material of the pipeline and valves' pup pieces shall be checked to ensure that the welding will be performed between correct materials.

DOI: 10.1201/9781003343318-12

- It is essential to mount the valve on the pipeline accurately and parallelly without any offset between the pup pieces and pipeline. This point is necessary to know that valves are not designed to eliminate misalignments between two pipeline pieces.
- The welder should carry out an initial accurate spot welding to verify the perfect alignment of the valve.
- It is better to perform welding on both sides of valves at the same time to reduce the tensions applied to valves by welding only one side.
- The preservatives that had been used during packing and preservation on both ends of the pipeline valves shall be cleaned and degreased by a suitable degreaser and a cloth before installation of valves.
- Double isolation and bleed (DIB) 2 type valves are unidirectional, as explained in Chapter 3, meaning that they have preferred installation direction. The flow direction is typically engraved on these valves' bodies or written on a tag plate securely fastened to the valve body with four rivets. Attention should be made to ensure that unidirectional pipeline valves are installed correctly according to the flow direction marked on them.
- The pipeline usually contains abrasive particles such as weld slags and sands that could damage the valve internals, so the pipeline must be thoroughly flushed before welding.
- The welder performing welding between the valves and pipeline shall be qualified based on relevant welding code such as ASME Boiler and pressure vessel code (BPVC) Sec. IV, titled "welding, brazing, and fusion qualifications," and the welder shall follow the welding procedure.
- The welder shall qualify welds between the valves and pipeline based on performing welds on rings with the same heat numbers and materials as the valves' pup pieces and the pipeline.
- The valves shall be in the open position during welding operation.
- The welding temperature shall be controlled during welding operation to prevent possible damages to the non-metallic seals installed inside the valves.
- Onsite personnel shall clean the pipeline and valve after welding.
- After welding completion, the valve's performance shall be checked by the operator by one closing and one opening. The operator shall not keep the valve in a half-open position and shall ensure that the inside of the valve is clean during operation.
- If any difficulties happen during operation, the valve manufacturer shall be contacted.
- An operator may remove the valve internals such as ball and seats to prevent damage during pre-commissioning and before start-up. Pre-commissioning refers to series of activities on the pipeline before start of operation to ensure that the pipeline is clean and can handle the production fluid without any leakage. In such a case, a removable and temporary sleeve is installed inside the valve since the valve internals are removed.
- The actuator is mounted on top of the valve when installing the valve to the pipeline. For a hydraulic actuator, the end stops installed on the actuator

shall be adjusted by an operator to provide the valve's accurate full opening and closing during operation. The operator can apply one cycle (opening and closing) to ensure the intersection between the ball and seats.

12.2 MAINTENANCE

Generally, pipeline valves are maintenance-free and are designed to last for a long period of time such as 20 to 30 years. However, pipeline valves are installed on the pipelines to control the production of the final oil and gas products, and high safety and reliability are expected from these valves. Thus, the valves should be inspected regularly by inspectors to ensure that there are no internal leakages from the seats of the valves. In addition, there should be no external leakages from the stem or any other connection to the valve. By injecting sealants into sealant injection ports, injection sealants can prevent leakages from seats and stems. Maintenance personnel may be able to perform elementary repairs to prevent external leakage without dismantling the valve in some cases. An external leak that cannot be stopped by elementary repairs can be prevented by disassembling the valve and replacing worn parts. It is imperative that the valve manufacturer provides the purchaser (end user) with a document titled Spare Parts Interchangeability Record (SPIR) which lists all of the spare parts available for the valve that may be used as replacements for damaged valve components during maintenance.

12.2.1 Emergency Sealant Injection

Using sealant injection to repair worn stem or seat seals is considered an emergency and temporary maintenance procedure. This means that the wearing parts will be replaced as part of a scheduled complete maintenance program. Manufacturers of valves do not usually recommend regular and routine injections of sealant. The sealant is typically a grease used to prevent or reduce leaks from seals installed around the stem or between the seats and the ball. Seals may be damaged by a variety of causes, such as particles, wear, and erosion. The maintenance operator uses a gun or a pump that provides sufficient pressure for sealant injection. A valve manufacturer recommended sealant type DESCO 800 or equivalent in one offshore project. DESCO 800 has enhanced resistance to hydrocarbons with an excellent metal adhesion property and resistance to water and oxidation. The following section describes the sealant injection procedures on the seats and stems.

12.2.1.1 Sealant Injection on Seats

Figure 12.1 illustrates the schematic of the sealant injection ports or fittings on the stem and seats of the valve. Item #41 is a cap or plug used to blind the sealant injection arrangement, item #43.

The operator shall perform the following steps for injection of sealants on seats:

- Put the valve in the closed position.
- Remove the cap of sealant injection fitting (item #41).
- Connect the grease pump to the seats' sealant injection fittings.

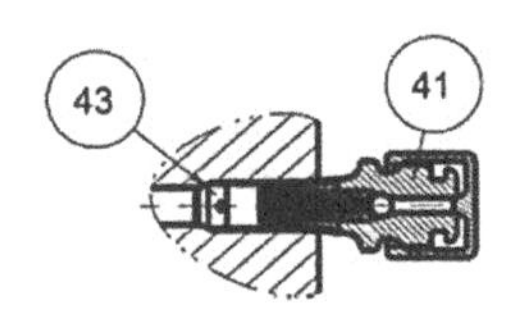

FIGURE 12.1 Sealant injection port or fitting on the stem and seats. (Photo by the author.)

- Pump in the sealant in a way that injection pressure shall not exceed the valve maximum operating pressure.
- Operate the valve just 10° from closed to open position and then close the valve again.
- Repeat pumping the sealant.
- Disconnect the sealant pump and reinstall the cap on the sealant injection fitting.
- Apply closure or seat test on the repaired seat of the valve and repeat the steps mentioned above if the valve has still unacceptable leakage rate through the seat.

12.2.1.2 Sealant Injection on Stem

- Remove the cap of sealant injection fitting (item #41).
- Connect the pump to the stem or shaft sealant fitting typically installed on the valve's body.
- Pump the sealant, and the injection pressure should be at a maximum of 1.5 times the valve operating pressure.
- Operate the valve just 10° from closed to open position and then close the valve again.
- Apply the leak test on the damaged stem seal and verify if the repaired seal can provide an acceptable sealing after the sealant injection.
- If the seal still leaks and the leakage rate is not acceptable, repeat the abovementioned steps until the seal can provide an acceptable leak rate.
- Disconnect the pump and install the cap on the grease injection fitting.

12.2.2 Minor External Leakage Repairs

A simple visual inspection is sufficient to reveal the defects that lead to external leakage in most cases. The external leakage can occur in at least one of the following connections, refer to Figure 12.2:

- Leakage from the body (item #1) and the bonnet (item #2) contact areas through seals (items # 499, 500, and 502)
- Leakage from sealant injection fittings (items #41 and 43)
- Leakage from the vent and drain flanges seals (items #537 and 538)
- Leakage from the stem and through seals (items # 506, 507, 508, and 510)
- Leakage from the flushing ports and modular valves (item # 44) and the gasket (item #546) between valves and the body of the pipeline valve (item #1)
- Leakage from seat retraction connections

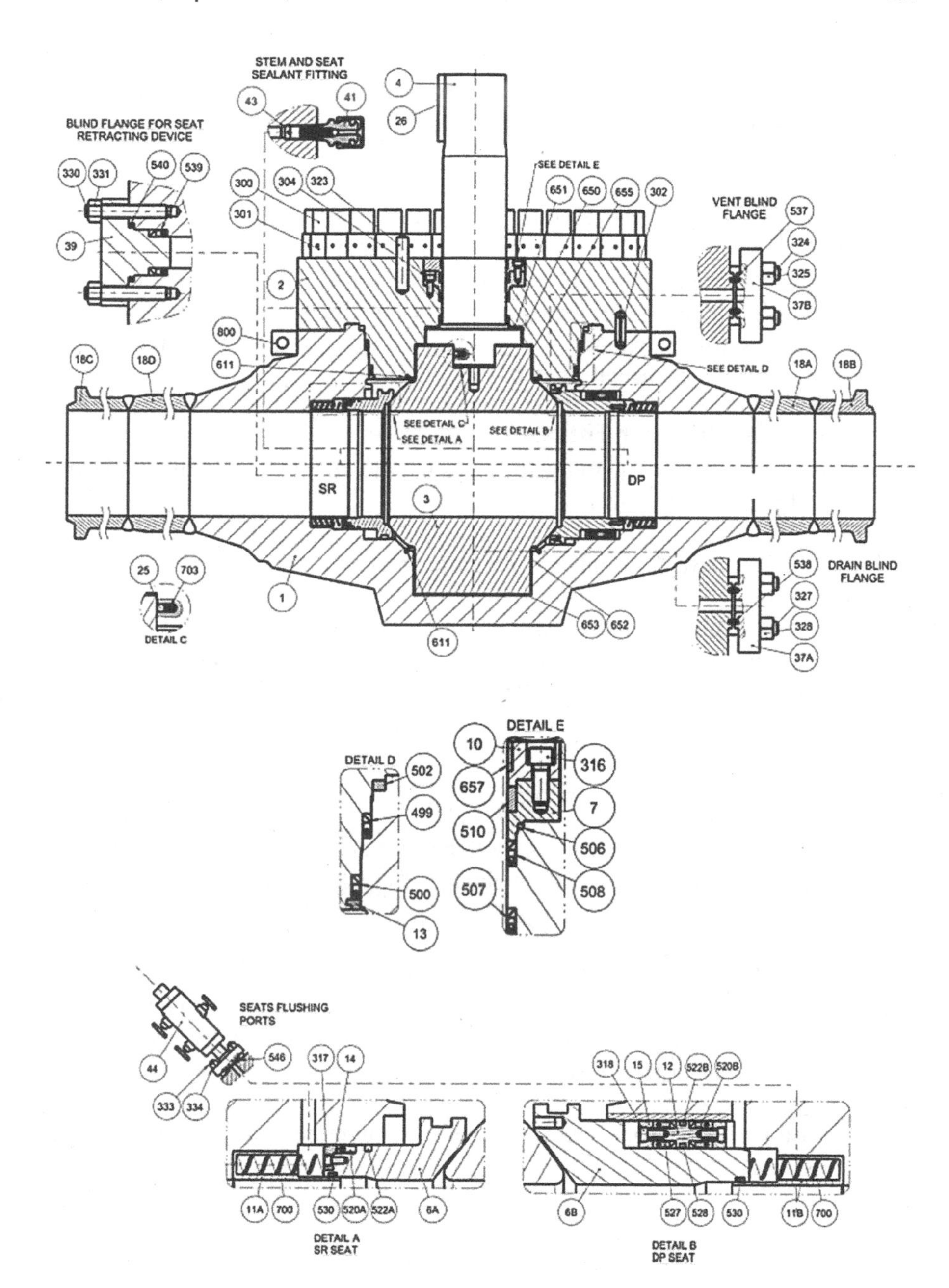

FIGURE 12.2 Pipeline valve cross-section drawing. (Photo by the author.)

Note 1
An explanation is provided in this chapter regarding the seat retraction tools used to disassemble the ball from the valve during maintenance.

Note 2
All repair activities to prevent external leakage can be conducted when the valve is installed on the pipeline. However, it is necessary for maintenance to isolate the valve from the pressure inside the pipeline. Before starting the repair, it is also necessary to release the pressure inside the valve cavity. The pipeline valves in gas service have vent blind flanges, while pipeline valves in oil service have drain blind flanges which allow the pressure inside the valves to be evacuated.

12.2.2.1 Sealant Injection Fitting

The operator shall perform the following tasks in order to repair the sealant injection fittings, as illustrated in Figure 12.3:

- Release the valve from internal pressure through the vent or drain cavity connection.
- Make sure that there is no pressure in the injection fitting port.
- Remove the injection plug (item #41).
- Remove the mini check valve (item #43).
- Clean all the surfaces.
- Install a new mini check valve.
- Reinstall the injection fitting injection.
- Install a new injection plug (item #41).
- Tighten the injection fitting slightly to achieve an acceptable sealing.
- Pressurize the valve and check for any leak from the repair sealant injection.

12.2.2.2 Vent and Drain Blind Flanges

Vent and drain blind flanges are installed in the cavity of the valve, and an operator opens them to release the pressurized fluid. As shown in Figure 12.4, the operator must perform the following tasks in sequence to repair the vent and drain blind flanges:

- Release the valve from internal pressure through the vent or drain cavity connection.

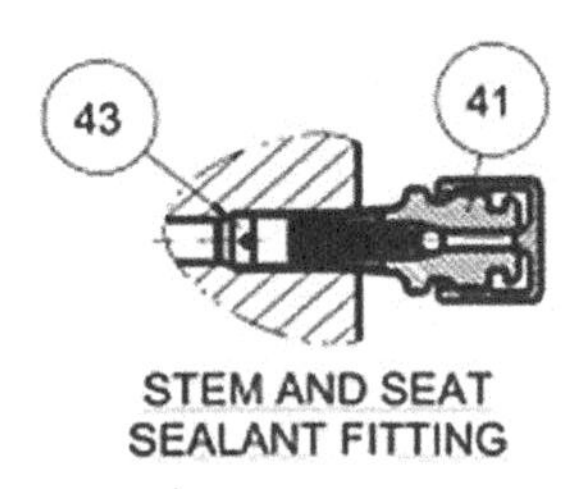

FIGURE 12.3 Injection fitting for the stem and seats. (Photo by the author.)

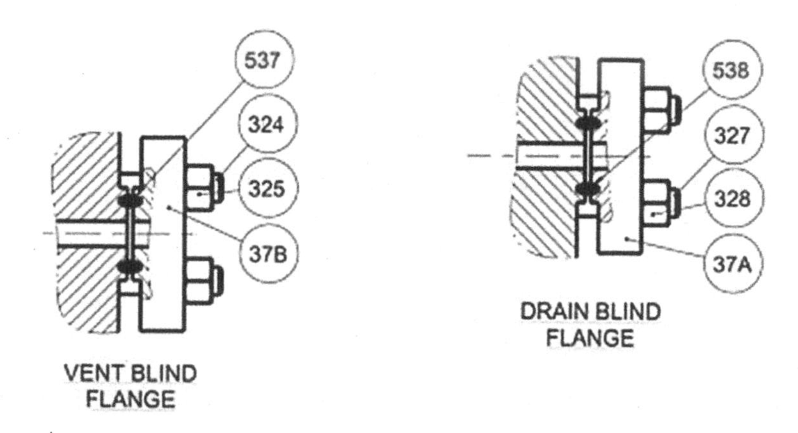

FIGURE 12.4 Vent and drain blind flanges. (Photo by the author.)

- Unscrew and remove the nuts (item #325 nuts for vent blind flange and item #328 nuts for drain blind flange).
- Remove the drain blind flange (item #37A) and vent blind flange (item #37B) by sliding them through drain stud bolts (item #327) and vent stud bolts (item #324), respectively.
- Check the roughness on the vent and drain flange sealing surfaces. The blind flanges for pipeline valves have a ring type joint (RTJ) face that should have a smooth gasket surface finish or a gasket groove finish surface, which is at a maximum of 1.6 μm equal to 63 μinch.
- If there is any corrosion product on the flange faces, use a clean emery cloth grade 500 or 1000 to remove the corrosion product and lubricate the flange faces afterward. An emery cloth-like sandpaper is a man-made abrasive that contains natural minerals.
- Install the RTJ gaskets (item #537 for the vent blind flange and item #538 for the drain blind flange) on the flange faces machined in the valve's body (item #1).
- Do not use any metallic tools for disassembling the flanges because it could damage the gaskets and flange faces.
- Reinstall vent and drain blind flanges by sliding them through the bolts.
- Reinstall and tighten the nuts according to the tightening torque provided in the installation, operation, and maintenance (IOM) manual supplied by the valve supplier.
- Pressurize the valve again and check if there are leakages from the vent and drain blind flanges.

12.2.2.3 Modular Valves on the Seat Flushing Ports

In Chapter 5, the function of seat flushing ports is described in detail. The entrapment of debris and particles inside valve seats is one cause of pipeline valve failure. It is

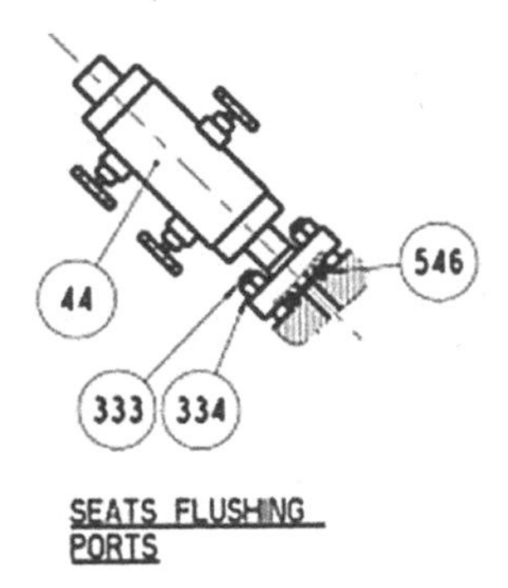

FIGURE 12.5 Seat flushing port arrangement including the modular valve. (Photo by the author.)

possible that crude oil contains waxes, which can deposit in seats and impair their functionality and tightness to the ball, resulting in seats leaking. One way to resolve this problem is to use flushing ports to inject some solvent or steam to dissolve the wax. The flushing ports are blinded with modular valves (see Figure 12.5) to provide double isolation between the internal fluid and the surrounding environment. In order to repair the modular valve connection, the operator should perform the following tasks in sequence:

- Unscrew and remove the nuts (item #334).
- Remove the blind flange by sliding it through stud bolts (item #333).
- Remove the RTJ gasket (item #546).
- Check the roughness on the modular valve and blind flange sealing surfaces. The flange faces should have a smooth gasket surface finish or a gasket groove finish surface, which is at a maximum of 1.6 μm equal to 63 μinch.
- If there is any corrosion product on the flange faces, use a clean emery cloth grade 500 or 1000 to remove the corrosion product and lubricate the flange faces afterward.
- Reinstall the blind flange by sliding it through the bolts.
- Reinstall and tighten the nuts according to the tightening torque provided in IOM manual supplied by the valve supplier.
- Pressurize the valve again and check if there are leakages from the seat flushing port.

12.2.3 Major External Leakage Repairs

The connection between the valve body and bonnet may require a significant repair. In this case, it is essential to remove the actuator as well as an adopter flange, also called coupling from the valve. Removal and reassembling of the actuator and adaptor flange as illustrated in Figure 12.6 is done through the following procedure:

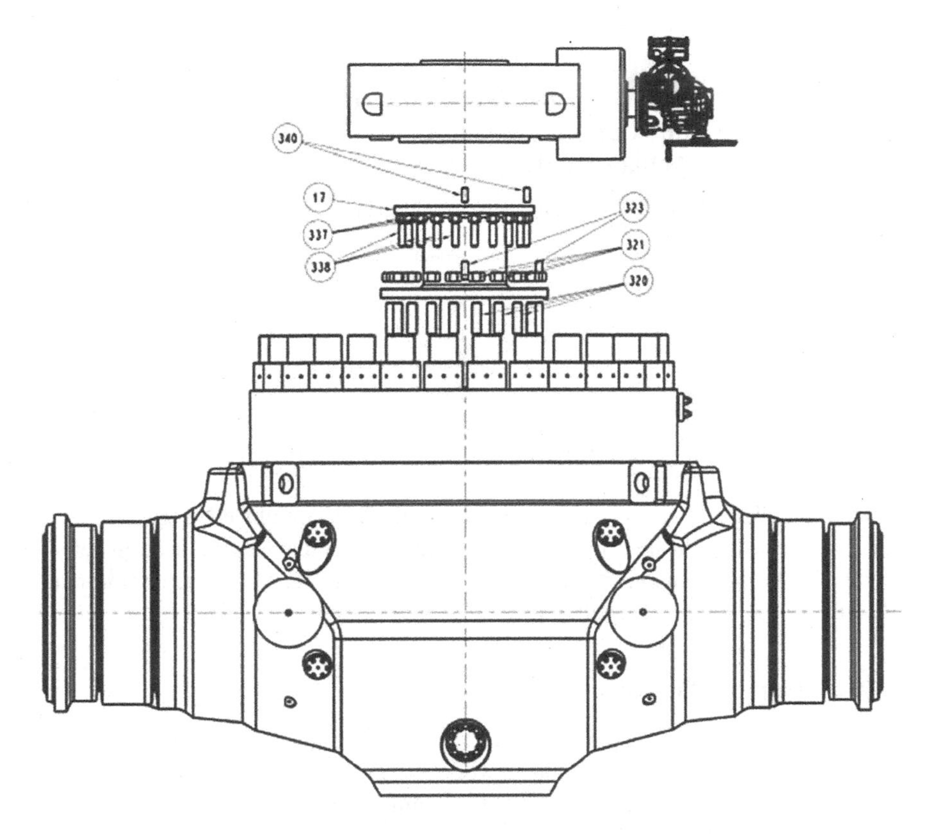

FIGURE 12.6 Removing the actuator and adaptor flange. (Photo by the author.)

- Unscrew and remove the nuts (item #338).
- Remove the actuator.
- Hook the adaptor flange extension (item #17) with adapted tools.
- Follow the procedure the other way around to reassemble the actuator and adaptor flange.
- Tighten all the nuts and screws according to the tightening procedure given in IOM.

Parts list for the actuator and adaptor flange connection to a 38" CL1500 pipeline ball valve is provided in Table 12.1.

The operator can separate the bonnet from the body by unscrewing the nuts (item #301). While removing the bonnet from the body, the operator must ensure that the stem is not removed. If there is any corrosion product on the body and bonnet, the maintenance personnel shall use a clean emery cloth grade 500 or 1000 to remove the corrosion product. The personnel can repair any damage to the stem seals and bearings after removing the bonnet and body. The other reason for external leakage from the body and bonnet gasket and stem seals could be the inappropriate roughness

TABLE 12.1
Parts List for the Connection of a 38" CL1500 Pipeline Ball Valve to an Actuator and Adaptor Flange

Item	Quantity	Designation
17	1	Adaptor flange extension
320	20	Stud bolt
321	20	Nut
323	2	Pin
337	20	Stud bolt
338	20	Nut
340	2	Pin

values of the metallic pockets and areas where the seals are placed. For example, the high roughness of stem areas in contact with stem seals increases the friction between the stem and seals, causing damages to seals and leakage. Thus, the operator must check the roughness of the sealing areas.

12.2.4 Internal Leakages

It is possible to perform a simple test to determine whether the valve is sufficiently tight during operation. The valve must be closed, and the vent or drain flange must be monitored for leaks. The vent or drain blind flanges can be monitored to show leakages from seats to the valve's cavity. It can be hazardous to monitor pressurized fluid or gas through drain or vent flanges. If the volume of observed leakage from the cavity exceeds the leakage rate of the seat, there is a source of internal leakage. If there is excessive leakage from the seat, the first step is to inject grease onto the seat through a connected sealant injection fitting. If the leakage rate cannot be reduced by grease injection, the seat(s) and the ball must be removed, inspected, repaired, or replaced. The operator shall operate handwheels installed on the valve's body that are connected to seat retraction devices to disconnect seats from the ball and easily remove the ball and seats from the valve. The seat retraction tool (see Figure 12.7) is a special tool that contains a blind flange (item #13) attached to the gearbox and handwheel (item #12) and a gasket sealing (item #14) between the flange and gearbox. The load applied by the operator to the handwheel is transferred to the seats through eccentric tools in blue color (item #11). Seat retraction tool is not always designed for pipeline valves. One disadvantage of seat retraction tools is to add extra holes on the valve's body and more possibility of external leakage from the valve. Figure 12.8 illustrates the assembly of four seat retraction tools on the body of a pipeline valve.

The operator should follow the procedure below to replace the stem, ball, and seats if the valve is installed horizontally with a vertical stem.

- Put the valve in the open position.
- Remove the actuator and adaptor flange according to the procedure explained before.

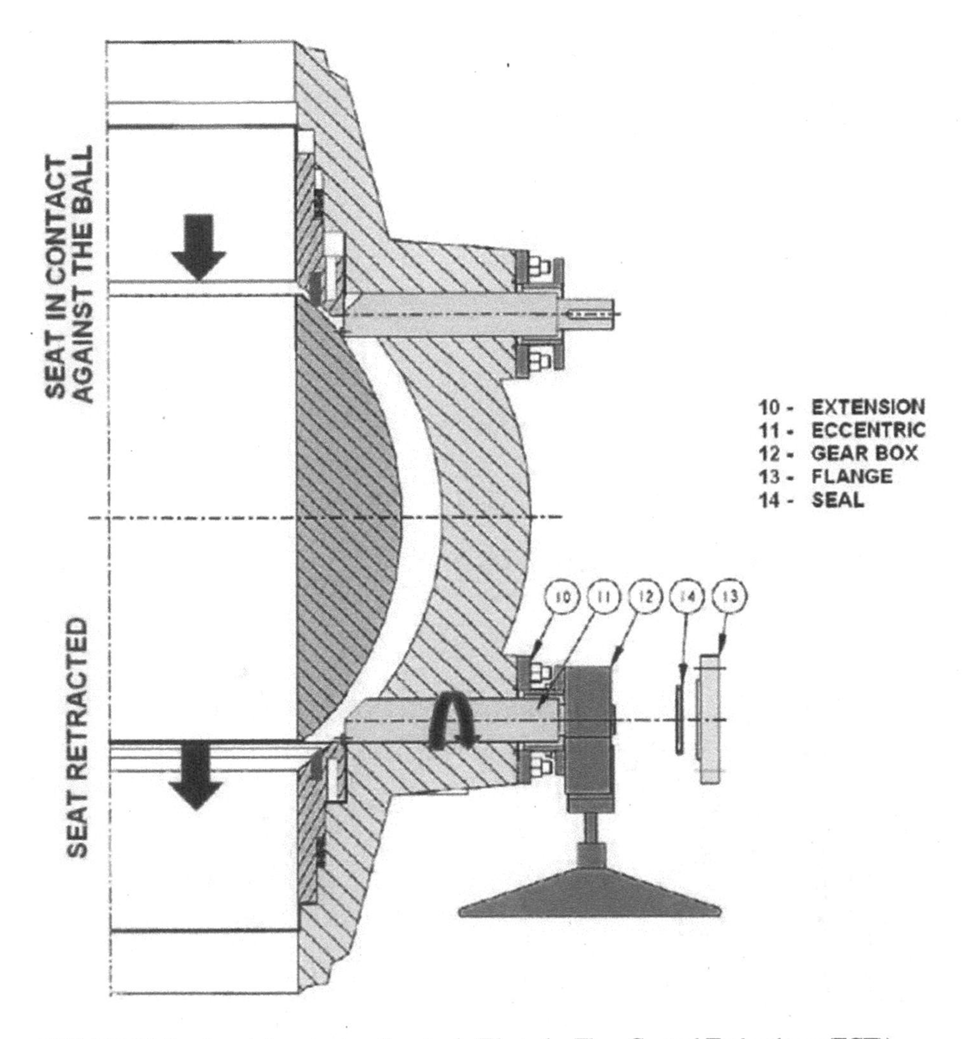

FIGURE 12.7 Special seat retraction tool. (Photo by Flow Control Technology (FCT)).

- Remove the bonnet.
- Remove stem seals and bearings and the stem.
- Remove the blind flanges installed on the seat retraction tools.
- Turn the handwheels on seat retraction tools to move the seats back from the ball.
- Remove the ball from the valve body through lifting lugs, as illustrated in Figure 12.9.
- Release two seats so that they return to their original positions.
- Remove the eccentrics.
- Remove the seats from the body.

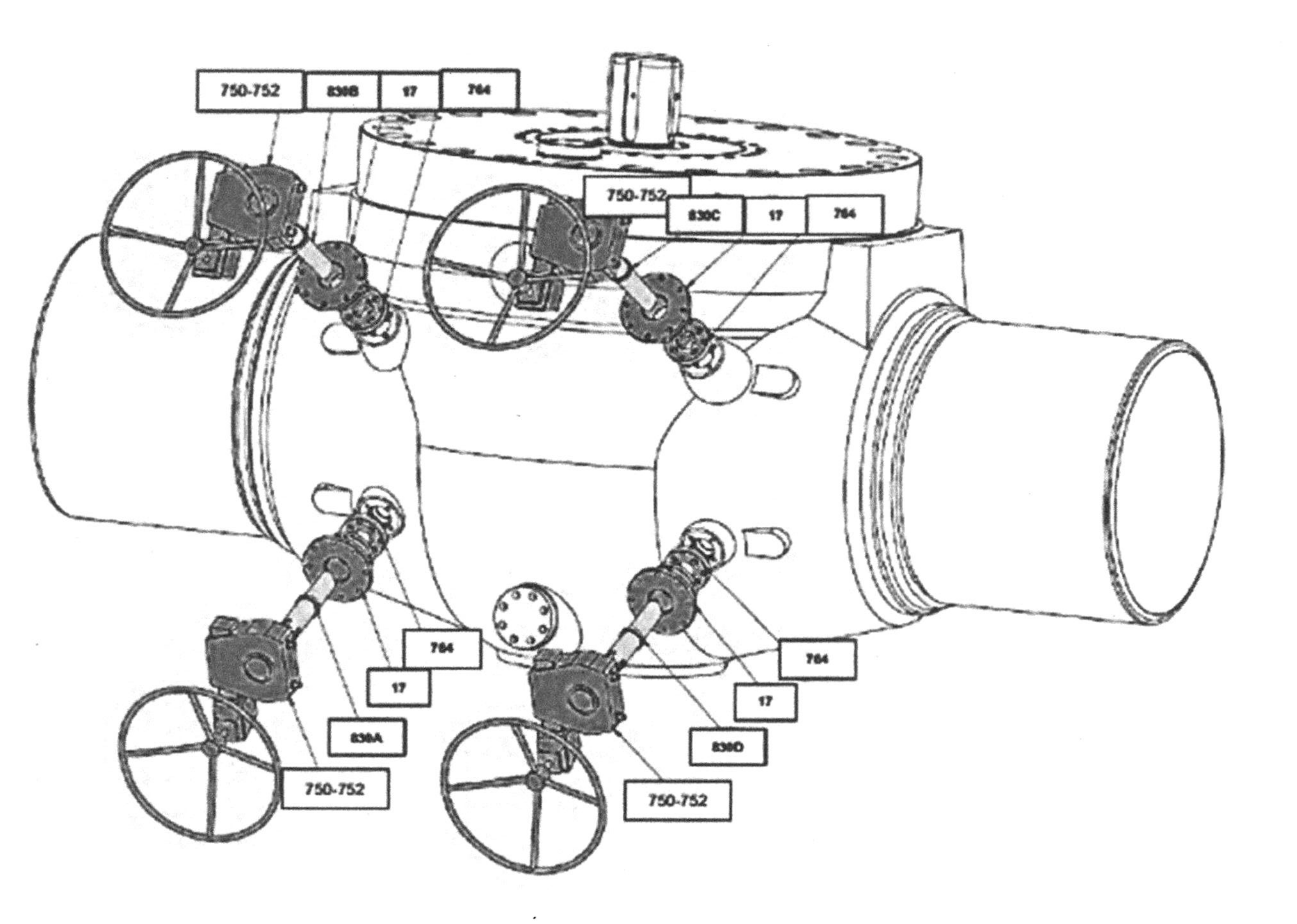

FIGURE 12.8 Four seat retraction arrangement assemblies on a pipeline valve. (Photo by Flow Control Technology (FCT)).

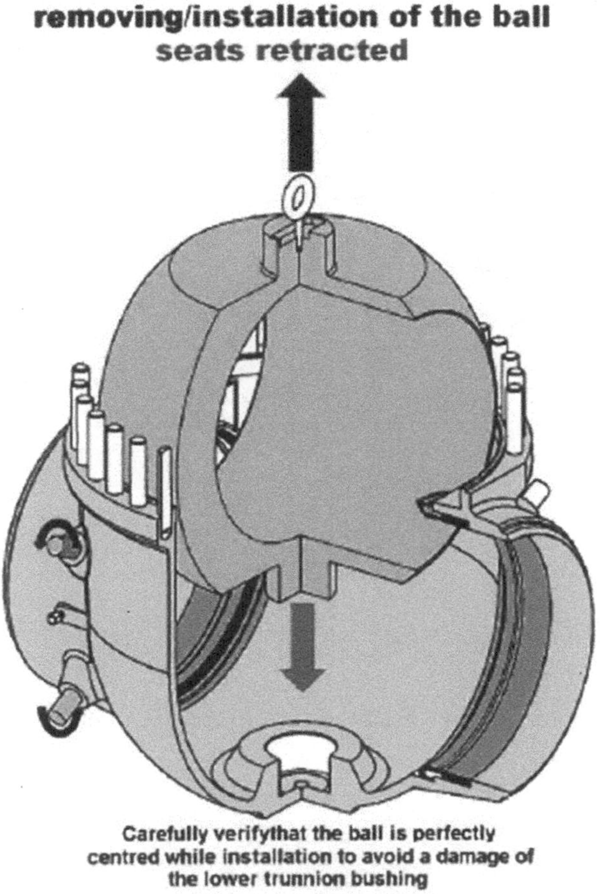

removing/installation of the ball
seats retracted

Carefully verifythat the ball is perfectly
centred while installation to avoid a damage of
the lower trunnion bushing

FIGURE 12.9 Removing the ball from the pipeline valve's body. (Photo by Flow Control Technology (FCT)).

The operator should follow the procedure below to replace the stem, ball, and seats if the valve is installed vertically with a horizontal stem.

- Remove the actuator and the adaptor coupling.
- Unscrew the bolts from the body and bonnet joint.
- Remove the bonnet, stem, and stem seals and bearings (see Figure 12.10).
- Use a special tool as illustrated in Figure 12.11 to remove the ball from the valve.
- Remove the seats.

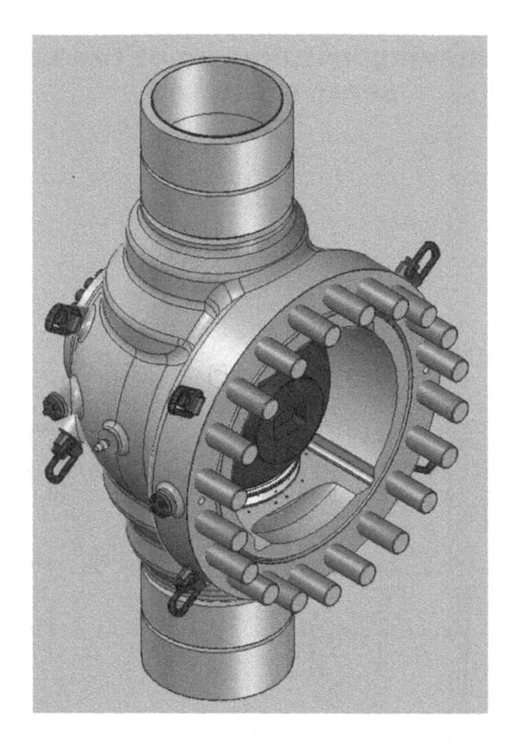

FIGURE 12.10 Removing the bonnet and stem from a vertically oriented valve with a special tool.

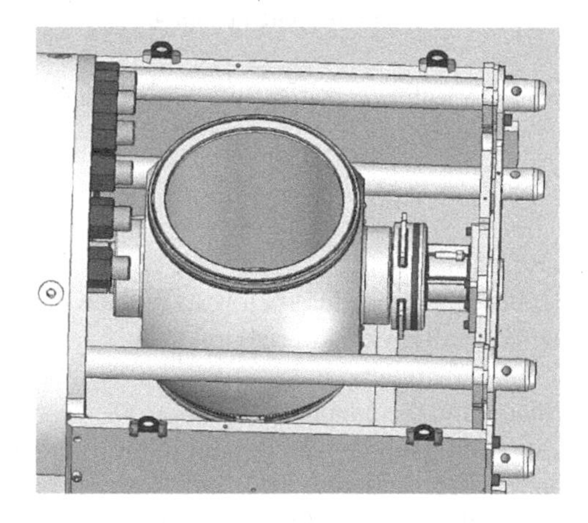

FIGURE 12.11 Ball removal from a vertically oriented valve by a special tool.

QUESTIONS & ANSWERS

1. Which sentence is correct regarding the storage and installation of pipeline valves?

    A. Pipeline valves can be stored for a long time outdoors.
    B. It is allowed to remove plywood sheets from the ends of a pipeline valve during storage.
    C. It is not required for welders of valves to the pipeline to be qualified as per the relevant welding code.
    D. The preservatives used during packing and preservation on both ends of pipeline valves shall be cleaned and degreased before the installation of valves.

    Answer) Option D is the correct answer. Pipeline valves are unpacked and cannot be stored outside for a long time, so option A is incorrect. Option B is incorrect since it is not permitted to remove plywood sheets from the ends of valves during storage. Option C is also incorrect as the welder of the valve to the pipe must be qualified.

2. During welding and operation, why is it important for onsite personnel to ensure the cleanliness of valves and the pipelines connected to them?

    A. Abrasive particles such as weld slags and sands prevent the operation of the valves.
    B. Abrasive particles can damage valves' internals, especially non-metallic or soft seals.
    C. Abrasive particles can affect the welding quality between valves and the pipeline.
    D. None of the above answers are correct.

    Answer) Option B is the correct answer.

3. Identify the correct sentences about sealant injection fittings.

    A. Maintenance personnel performs emergency and temporary repairs through sealant injection fittings.
    B. Sealant injection fittings are installed on one seat and the stem of pipeline valves.
    C. The sealant injected through sealant injection fittings is typically grease.
    D. The valves should be in the open position during sealant injection.
    E. The sealant injection is mainly employed to repair the seals used around the stem and between seats and the body of valves damaged due to wear, erosion, or particles.

    Answer) Options A, C, and E are correct sentences. Option B is incorrect since sealant injection ports or fittings are installed on both seats and the stem. Option D is wrong because the valve should be closed during sealant injection.

4. External leakages originating from which components?

   A. Vent and drain flanges
   B. Modular valves on seat flushing ports
   C. Body and bonnet connection
   D. All options are correct

   Answer) Option D is correct.

5. Which sentence is correct regarding seat retraction tools?

   A. The operator shall operate the handwheel on seat retraction tools if stem replacement is needed.
   B. Seat retraction tool design and installation on the valve's body does not have any disadvantage.
   C. The eccentric provides the contact point to the seat.
   D. Each seat requires one seat retraction tool for a pipeline valve in a 40" size.

   Answer) Option A is incorrect because seat retraction tools are not necessary to remove and repair the stem. It is incorrect to choose option B since seat retraction tools have one major disadvantage: they provide more holes on the valve body, increasing the likelihood of leakage. Option C is the correct answer. Option D is incorrect, as two seat retraction tools are required for a 40" size pipeline valve.

6. Find the wrong sentence regarding in-line maintenance for pipeline valves illustrated in Figure 12.12 equipped with seat retraction tools.

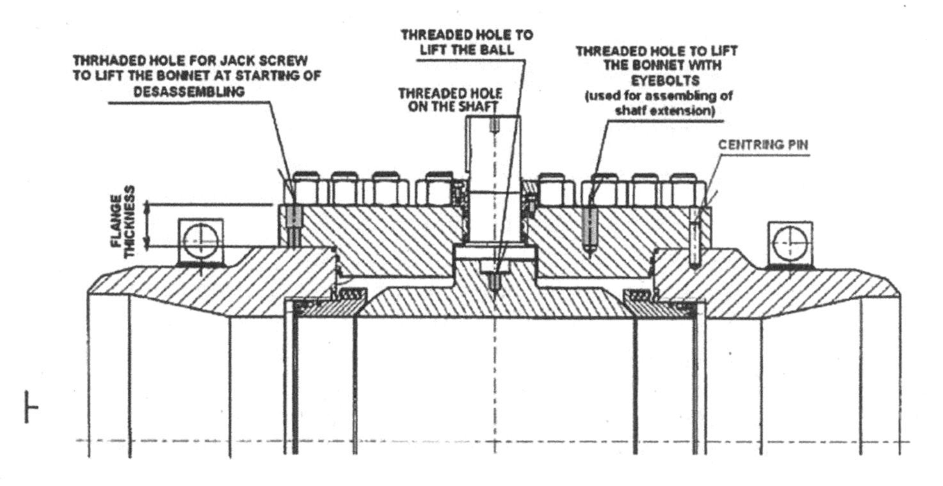

FIGURE 12.12 Schematic of a pipeline valve. (Credited to the author.)

   A. Seat retraction tool arrangements are installed on the valve's body.
   B. The eccentric tool moves the seats back from the ball.
   C. Lifting lugs are provided for bonnet, stem, and ball.
   D. The removal sequence in order is bonnet, stem, seats, and the ball.

Answer) Options A, B, and C are correct. Option D is incorrect because the removal sequence is the bonnet, stem, ball, and then seats.

7. Which statements are incorrect regarding Figure 12.13?

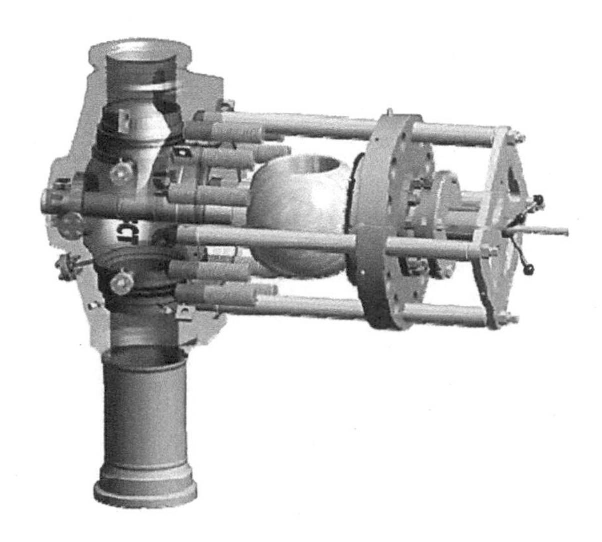

FIGURE 12.13 Ball removal from the valve.

A. The figure shows in-line maintenance for a pipeline ball valve.
B. The special tool used for ball removal is used to maintain pipeline valves installed on both horizontally and vertically oriented pipelines.
C. The valve is the top entry and welded to the pipeline.
D. All the above statements are correct.

Answer) Options A and C are correct. Option B is wrong because the special tool is only used to remove the ball of vertically oriented pipeline valves with the stem in a horizontal plane. Option D is incorrect because not all the statements are correct.

BIBLIOGRAPHY

1. American Society of Mechanical Engineers (ASME). (2019). *Boiler and pressure vessel code, Section IX: Welding and brazing qualifications*. New York, NY: ASME.
2. Sotoodeh K. (2021). *A practical guide to piping and valves for the oil and gas industry*. 1st edition. Austin, USA: Elsevier (Gulf Professional Publishing).

13 Safety and Reliability

13.1 INTRODUCTION TO SHUTDOWN SYSTEMS

Typically, pipeline valves are automated with either electrical or hydraulic actuators, as discussed in Chapter 8. In terms of safety, electrically actuated pipeline valves are not considered critical. On the other hand, the last valve installed on the pipeline has a hydraulic actuator with an emergency shutdown (ESD) function that is considered a ***safety-critical valve***. Safety-critical valves are those that are connected to either a process shutdown (PSD) or an ESD system. There are two types of shutdown systems for process systems: PSD and ESD. The PSD system minimizes or prevents the occurrence and consequences of process parameters like pressure, temperature, and level exceeding operating limits as part of plant and facility safeguarding systems. PSD systems can stop a part or all of the process or depressurize or blow down parts of it. The PSD valves are actuated valves connected to the basic process control systems that manage and control the process within accepted limits. In an ESD system, emergency conditions such as extreme or abnormal values of process parameters (e.g., high fluid flow in piping) or leakage or emission of hydrocarbons to the environment are minimized or prevented. Therefore, unlike PSD, ESD is a safety layer, not a process control layer, as shown in Figure 13.1. To prevent piping and equipment from being harmed by abnormal process conditions, ESD valves are actuated and closed in an emergency mode. Figure 13.1 illustrates the different layers of plant safety implementation and protection systems, including PSD and ESD. Sensors, valves, trip relays, and logic units for analyzing signals are typically included in both PSD and ESD systems. As a general rule, PSD and ESD valves in oil and gas projects are listed in ***safety analysis reports*** (SARs) and must meet ***safety integrity level (SIL)*** 2 or 3. The valves that provide isolation on pipelines used for export have the highest level of criticality. This means that failures of these valves could cause a large leak with a long duration and have severe consequences for safety. Thus, safety-critical pipeline valves should be leak tested during operation to determine their seal ability.

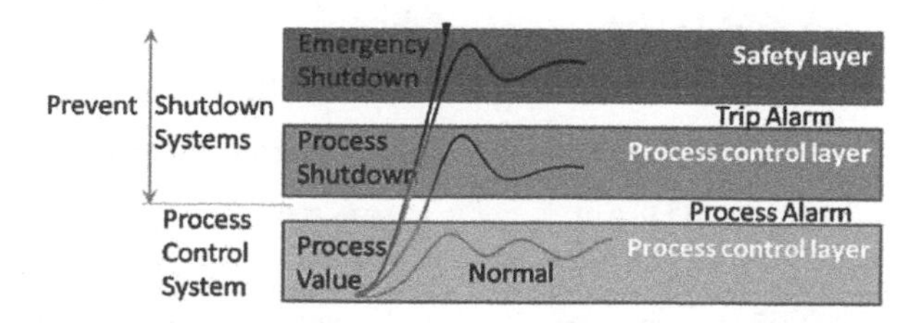

FIGURE 13.1 Plant safety implementation and protection with PSD and ESD layers.

DOI: 10.1201/9781003343318-13

13.2 SAFETY INTEGRITY LEVEL

SIL is a term attached to international standards such as IEC 61508 that provides suppliers and end-users with a common framework to design products and systems for safety-related applications. The International Electrotechnical Commission has published IEC 61508 as an international standard for electrical, electronic, and programmable electronic safety-related systems. SIL provides a scientific, numerical approach to specifying and designing safety systems to enable quantification of failure risk. The SIL provides a scientific and numerical approach to defining and evaluating safety systems, which quantifies their risk of failure. SIL levels range from 1 to 4, as shown in Table 13.1. There is the least probability of failure or probability of failure on demand (PFD) with SIL 4 and the highest reliability. PFD is a measure of a safety function's effectiveness. The likelihood that the safety function will not work when needed is represented by this value. As a result, SIL1 has the highest likelihood of failure and the lowest level of safety. A minimum of SIL 2 or 3 is typically required for ESD valves, as explained earlier.

Hydraulically operated ESD pipeline valves consist of three parts: valve actuator, valve control panel, and valve control system. In SIL analysis, we will use the 1oo1 architecture since these three components are considered a single component. According to Figure 13.2, this architecture consists of a single channel in which any failure of any part compromises the safe operation of the whole system.

Failure rates (probabilities) are categorized into two types: safe failure (λ_S) or dangerous failure (λ_D) rates. Dangerous failure rate relates to failures that will prevent achieving the required SIL. Safe failures put the safety function in its safe state, such as the ESD of a valve in case of failure. It is noticeable that SIL addresses only dangerous failures (λ_D). The dangerous failures can be either detected or undetected. The possibility of detected dangerous failures and undetected dangerous failures are shown as λ_{DD} and λ_{DU}. Equation 13.1 shows the relationship between λ_D, λ_{DD}, and λ_{DU}. Failure modes classification is shown in Figure 13.3.

Equation 13.1 Relationship between dangerous failure rate and dangerous detected and dangerous undetected failure rates according to IEC 61508

$$\lambda_D = \lambda_{DU} + \lambda_{DD}$$

TABLE 13.1
SIL Levels According to IEC 61508.

SIL	PFD	Risk Reduction Factor
SIL 1	0.1 to 0.01	10 to 100
SIL 2	0.01 to 0.001	100 to 1000
SIL 3	0.001 to 0.0001	1000 to10,000
SIL 4	0.0001 to 0.00001	10,000 to 10,0000

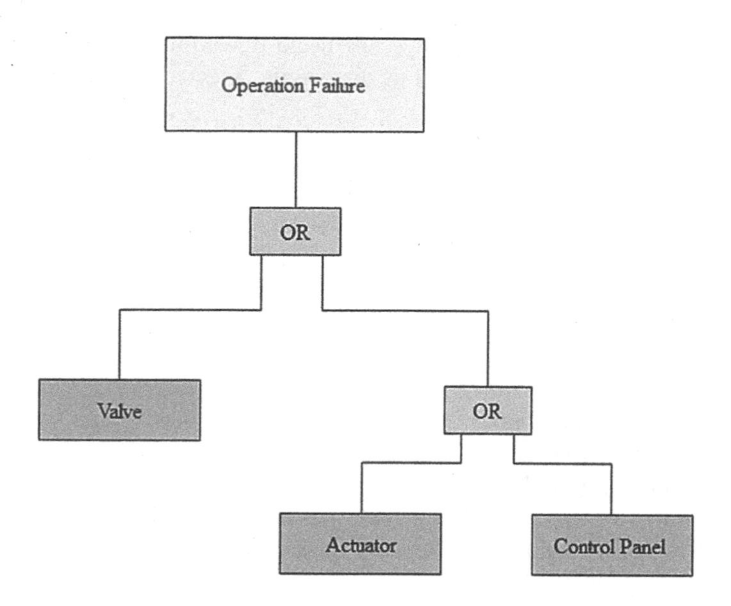

FIGURE 13.2 Fault tree for the ESD pipeline valve with hydraulic actuator.

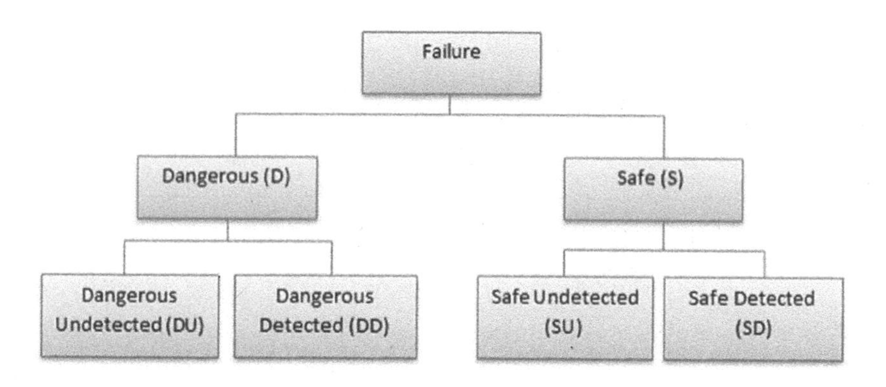

FIGURE 13.3 Failure modes classification.

The mean downtime of a system due to both dangerous detected and undetected failures $\mathbf{t}_{CE}$ includes the mean downtime of the system due to dangerous undetected failures $\mathbf{t}_{c1}$ plus the mean downtime of the system due to dangerous detected failures $\mathbf{t}_{c2}$ as shown in Figure 13.4.

$\mathbf{t}_{c1}$ and $\mathbf{t}_{c2}$ are calculated using Equations 13.2 and 13.3:

Equations 13.2 & 13.3 Calculation of the mean downtime of the system due to dangerous undetected failures as well as dangerous detected failures, refer to IEC 61508

$$\mathbf{t}_{c1} = \frac{T_1}{2} + \text{MRT} \tag{13.2}$$

$$\mathbf{t}_{c2} = \text{MTTR} \tag{13.3}$$

where:

T_1: proof test interval (hour)
MTTR: mean time to restoration (hour)
MRT: mean repair time (hour)
MTTR = time to detect the failure + time spent before starting the repair + the effective time to complete the repair + the time before the component is put back into operation (13.4)
MRT = *MTTR* – time to detect the failure (13.5)

Figure 13.4 shows the reliability block diagram for architecture 1oo1:

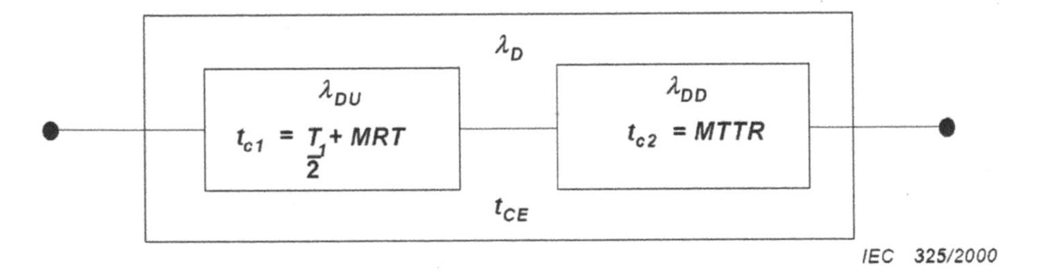

FIGURE 13.4 1oo1 reliability block diagram.

Placing values of $\mathbf{t}_{c1}$ and $\mathbf{t}_{c2}$ from Equations 13.2 and 13.3 into Equation 13.4:

$$\mathbf{t}_{CE} = \frac{\lambda_{DU}}{\lambda_D}\left(\frac{T_1}{2} + \text{MRT}\right) + \frac{\lambda_{DD}}{\lambda_D}\text{MTTR} \tag{13.4}$$

The number of dangerous failures detected by automatic online diagnostic tests is called diagnostic coverage (DC). DC is the detected dangerous failure rate divided by the total rate of the dangerous failures, using Equations 13.5, 13.6, and 13.7:

$$DC = \frac{\lambda_{DD}}{\lambda_D} \tag{13.5}$$

Therefore,

$$\lambda_{DD} = \lambda_D \times \text{DC} \tag{13.6}$$

$$\lambda_{DU} = \lambda_D - \lambda_{DD} = \lambda_D - (\lambda_D \times \text{DC}) \rightarrow \lambda_{DU} = \lambda_D (1 - \text{DC}) \tag{13.7}$$

The average PFD (PFD_G) is calculated using Equations 13.8 and 13.9:

$$PFD_G = (\lambda_{DU} + \lambda_{DD})\mathbf{t}_{CE} \tag{13.8}$$

$$\rightarrow PFD_G = \lambda_D \text{ x } \mathbf{t}_{CE} = \lambda_{DU}\left(\frac{T_1}{2} + \text{MRT}\right) + \lambda_{DD} \times \text{MTTR} \tag{13.9}$$

The other important safety parameter is called safe failure fraction (SFF), which verifies the suitability of a device for usage in a system for a particular SIL. On the basis of Equation (13.10, the SFF is calculated as the ratio of the average safe failure rate plus the average dangerous detection rate to the average total failure rate:

$$\text{SFF} = \left(\frac{\lambda_S + \lambda_{DD}}{\lambda_S + \lambda_D}\right) = \left(\frac{\lambda_{TOTAL} - \lambda_{DU}}{\lambda_{TOTAL}}\right) \tag{13.10}$$

SFF and SIL can be correlated based on a table for different hardware fault tolerance levels. Hardware fault tolerance can be N = 0, 1, or 2 and can either be type A or B. Type B provides a higher level of safety for more complex diagrams. Type A is suitable for a single block diagram and a simple 1oo1 architecture, which is why type A has been selected for the ESD pipeline valve. In Table 13.2, the maximum allowable SIL for a safety function implemented by a type A safety is illustrated for the different hardware types. A hardware fault tolerance of N indicates that an N+ 1 fault could result in the loss of safety functions. Since architecture type 1oo1 has been selected in this chapter, which means that failure of any component may cause failure

TABLE 13.2
Correlation Between SIL and SFF for Hardware Fault Tolerance – Type A (IEC)

Safe Failure Fraction (SFF)	Hardware Fault Tolerance – Type A		
	0	1	2
SFF < 60%	SIL1	SIL2	SIL3
60% < SFF < 90%	SIL2	SIL3	SIL4
90% < SFF < 99%	SIL3	SIL4	SIL4
SFF ≥ 99%	SIL3	SIL4	SIL4

of the entire system, the hardware fault tolerance value has been set to 0 in order to achieve the highest possible level of safety and reliability.

13.2.1 SIL Case Study for a Pipeline Valve

A 20" hydraulically actuated ball valve (see Figure 13.5) is designed to be installed on an offshore gas export pipeline that transfers gas to land. In a low-temperature carbon steel material ASTM A352 LCC, the valve has a pressure class 1500 equal to 258.6 at ambient temperatures (38°C) with an ESD function. Normally, the ball valve is open; however, in the event of a failure in the pipeline, it will be closed to halt production and allow for pipeline maintenance. Table 13.3 illustrates various failure modes for valves, actuators, and control panels, in addition to failure effects, the possibility of detecting defects through test, and related failure probabilities.

Table 13.4 summarizes the failure modes, such as probabilities of safe failures λ_S, dangerous detected failures λ_{DD},and dangerous undetected failures λ_{DU}. The valve, actuator, and control panel manufacturers have provided the data in Tables 13.3 and 13.4. This case study is being implemented to prove the valve, actuator, and control panel together as a single unit meet the SIL2 requirements of the client.

The next step is to calculate $\mathbf{t}_{CE}$ for the whole system using Equation 13.4.

T_1: proof test interval (hour) is 1.114 year, equal to 9760 hours.
MTR: maintenance repair time = 24 hours
Mean repair time (MRT) = MTTR – time to detect the failure

FIGURE 13.5 A 20" CL1500 ball valve with a hydraulic actuator will be installed on an export pipeline. (Photo by the author.)

TABLE 13.3
Failure Modes and Effects and Failure Probabilities for a 20" CL1500 ESD Ball Valve on a Gas Export Pipeline

	Failure Mode	Effect	Test Strategy
Valve (20" top entry ball valve class 1500)			
$\lambda_{DD} = 3 \times E^{-8}$ $\lambda_{DU} = 0.8 \times E^{-8}$	*Valve packing is loose (dangerous)*	*Leakage from the valve packing to the atmosphere*	*Testing the valve through full or partial stroke*
$\lambda_{DD} = 2 \times E^{-8}$ $\lambda_{DU} = 1.8 \times E^{-8}$	*Valve stem sticks (dangerous)*	*Valve fails to close*	*Testing the valve through full or partial stroke*
$\lambda_{DD} = 1.9 \times E^{-8}$ $\lambda_{DU} = 0.7 \times E^{-8}$	*Valve seat is scratched (dangerous)*	*Valve fails to seal off*	*Testing the valve through full stroke or leak test*
Actuator			
$\lambda_{DD} = 0$ $\lambda_{DU} = 0.7 \times E^{-8}$	*Actuator sizing is insufficient*	*Valve fails to close*	*Not possible to detect with test*
$\lambda_{DD} = 2.30 \times E^{-8}$ $\lambda_{DU} = 1.00 \times E^{-8}$	*The spring of the actuator has been broken*	*Valve fails to close*	*Testing the valve through full stroke test*
Control panel			
$\lambda_{DD} = 0$ $\lambda_{DU} = 1.3 \times E^{-7}$	*Wrong control panel selection*	*System malfunction*	*Not possible to detect with test*
$\lambda_s = 1.3 \times E^{-9}$	*Sharp edges in the control panel*	*Possible damage to the operator*	*This type of failure does not affect safety instrumented function(SIF) and is considered as a safe failure*

TABLE 13.4
Summarizing the Failure Probabilities for the 20" Pipeline Ball Valve

Component	λ_s	λ_D	λ_{DU}	λ_{DD}
Top entry 20" class 1500 ball valve	0	$1.02 \times E^{-7}$	$3.3 \times E^{-8}$	$6.9 \times E^{-8}$
Actuator	0	$4 \times E^{-8}$	$1.7 \times E^{-8}$	$2.3 \times E^{-8}$
Control panel	$1.3 \times E^{-9}$	$1.3 \times E^{-7}$	$1.3 \times E^{-7}$	0
Total values	$1.3 \times E^{-9}$	$2.72 \times E^{-7}$	$1.8 \times E^{-7}$	$9.2 \times E^{-8}$

Time to detect the failure is zero since there are automatic detectors, called valve watches, installed on the valve. Therefore, MTTR=MTR=24 hours

$$\mathbf{t}_{CE} = \frac{\lambda_{DU}}{\lambda_D}\left(\frac{T_1}{2} + \text{MRT}\right) + \frac{\lambda_{DD}}{\lambda_D}\text{MTTR}$$

$$\mathbf{t}_{CE} = \frac{1.8\times E^{-7}}{2.72\times E^{-7}}\left(\frac{9760}{2}+24\right)+\frac{9.2\times E^{-8}}{2.72\times E^{-7}}\times 24 = 3245+8.12 = 3253\,\text{hours}$$

Now, it is possible to calculate PFD_G using Equation 13.8:

$$PFD_G = \lambda_D \times \mathbf{t}_{CE} = 2.72\times E^{-7}\times 3253 = 8.85\times E^{-4}$$

The next step is to calculate the SFF for the whole system using Equation (13.10):

$$\text{SFF} = \left(\frac{\lambda_S+\lambda_{DD}}{\lambda_S+\lambda_D}\right) = \left(\frac{1.3\times E^{-9}+9.2\times E^{-8}}{1.3\times E^{-9}+2.72\times E^{-7}}\right) = 34.13\%$$

According to Table 13.2, the SFF is less than 60%, so the safety level of the system complies with SIL1 and does not meet the project requirements. We need to improve the safety and reliability of the system. Assume that the failure probabilities of the valve in this case study are changed to those given in Table 13.5.

$$\mathbf{t}_{CE} = \frac{\lambda_{DU}}{\lambda_D}\left(\frac{T_1}{2}+\text{MRT}\right)+\frac{\lambda_{DD}}{\lambda_D}\,\text{MTTR}$$

$$\mathbf{t}_{CE} = \frac{5.23\text{E}-07}{6.06\text{E}-07}\left(\frac{9760}{2}+24\right)+\frac{8.34\text{E}-08}{6.06\text{E}-07}\times 24 = 4235.30$$

Now, it is possible to calculate PFD_G using Equation (13.10):

$$PFD_G = \lambda_D \times \mathbf{t}_{CE} = 6.06\text{E}-07\times 4235.30 = 2.56\text{E}-03$$

The next step is to calculate the SFF using Equation 13.10:

$$\text{SFF} = \left(\frac{\lambda_S+\lambda_{DD}}{\lambda_S+\lambda_D}\right) = \left(\frac{1.05\text{E}-06+8.34\text{E}-08}{1.05\text{E}-06+6.06\text{E}-07}\right) = 68.44\%$$

TABLE 13.5
A Summary of the Updated Failure Probabilities for the 20" Pipeline Valve

Component	λ_s	λ_D	λ_{DU}	λ_{DD}
TE 20" class 1500 ball valve	0	$8.55\times E^{-8}$	$2.57\times E^{-8}$	$5.99\times E^{-8}$
Actuator	0	$2.58\times E^{-8}$	$2.3\times E^{-9}$	$2.35\times E^{-8}$
Control panel	$1.05\times E^{-6}$	$4.95\times E^{-7}$	$4.95\times E^{-7}$	0
Total values	$1.05\times E^{-6}$	$6.06\times E^{-7}$	$5.23\times E^{-7}$	$8.34\times E^{-8}$

As shown in Table 13.2, the SFF is between 60% and 90%, so the safety level of the system complies with SIL2. Using these steps, calculating the SFF for each system component separately, including the valve, actuator, and control panel, provides an SFF of 69.98%, 91.09%, and 68%. Therefore, every single component of the system offers an SIL of at least 2. The following section explains another safety and reliability method called failure mode and effect analysis (FMEA).

13.3 FAILURE MODE AND EFFECT ANALYSIS

FMEA is used to identify and eliminate known and potential problems and errors from a system before it is delivered to the customer. The tool is widely used for reliability analysis. A major benefit of this system is that it improves safety, reliability, and reduces the possibility of failure, thereby increasing customer satisfaction. Human injuries and adverse environmental impacts can be minimized through failure reduction opportunities that improve HSE (health, safety, and environment). This is the most common risk analysis approach currently utilized. According to history, FMEA was developed by the American military at the end of the 1940s and implemented in the 1950s to design flight control systems. The following example shows how to use FMEA for pipeline valves.

13.3.1 FMEA Case Study for Pipeline Valves

Figure 13.6 shows the flowchart and steps for FMEA analysis that are used as a basis for FMEA research methodologies. As pipeline valves are the most critical components in an offshore platform or production unit, they are selected for FMEA analysis in Step 1. The second step in the flowchart identifies the failure modes of pipeline valves. We collect failure data in Step 2 based on actual failures of pipeline valves in the field as well as risk identification by pipeline valve manufacturers. Step 3 is to select a failure mode for analysis. Step 4 is to explain the consequences of each failure. The next step is to evaluate the severity and frequency or probability of occurrence for each of the failure modes identified. Equation 13.11 can now be used to calculate the risk priority number (RPN):

Equation 13.11 Risk priority number calculation

$$\text{Risk priority number}\left(\text{RPN}\right) = \text{Severity x Probability of occurrence}$$

The severity of the potential failure mode effect is determined by a scale from 1 to 10. The higher the severity, the greater the risk of failure mode occurrence. According to Table 13.6, severity levels range from 1 to 10.

According to Table 13.7, the probability of occurrence has been defined and classified. Each potential failure risk is rated between 1 and 10.

Table 13.8 summarizes the failure modes and consequences, as well as possible causes of failure and RPNs.

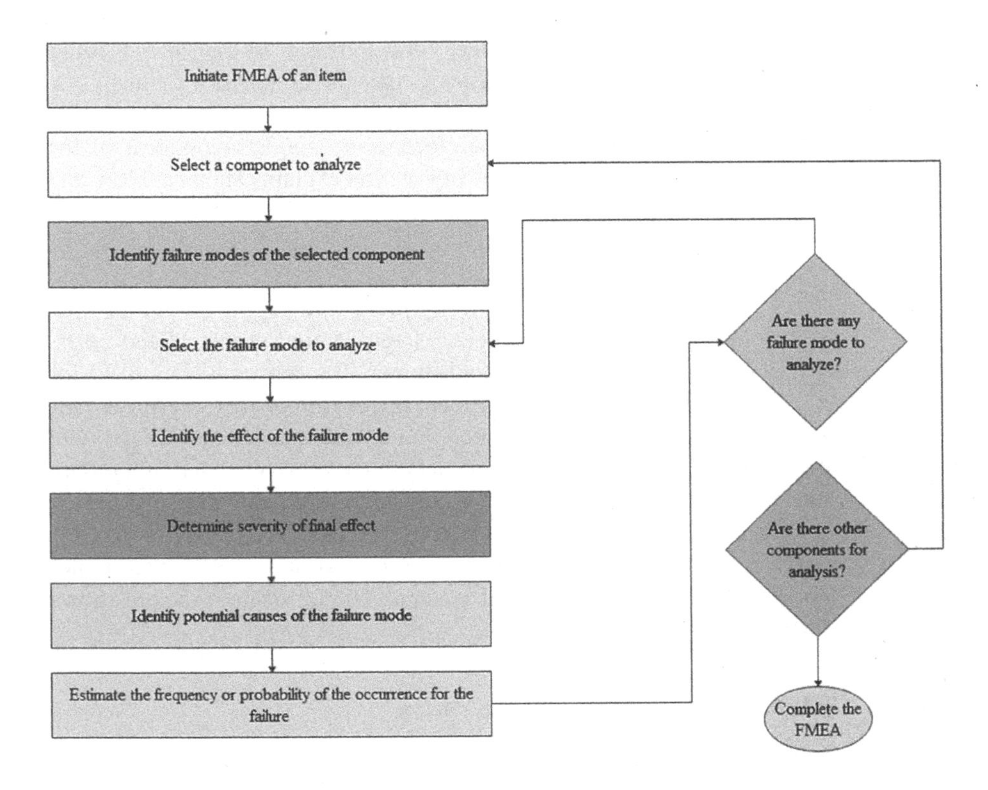

FIGURE 13.6 FMEA procedure chart. (Credited to the author.)

TABLE 13.6
FMEA Severity Effect Determination Table

Rating	Effect	Severity Effect
10	Hazardous without warning	Very high severity without warning
9	Hazardous with warning	Very high severity with warning
8	Very high	Destructive and unsafe failure
7	High	System is inoperable due to major equipment damage
6	Moderate	System is inoperable due to minor damage
5	Low	System is inoperable without damage
4	Very low	High degradation of performance
3	Minor	Less performance degradation
2	Very minor	Minimal system operability
1	None	No effect

TABLE 13.7
Possibility of Failure Occurrence

Rating	Occurrence Probability	Minimum Probability Percentage
10	Extremely high	50%
9	Very high	33%
8	Very high	10%–15%
7	High	5%
6	Marginal	1%
5	Marginal	0.25%
4	Unlikely	0.05%
3	Low	0.007%
2	Very low	0.0007%
1	Remote	0.000007%

TABLE 13.8
FMEA Analysis of Pipeline Valves

Valve/ Actuator Component	Potential Failure Mode	Potential Effect of Failure	Severity (S)	Cause of Failure	Occurrence	RPN
Valve body and seat	Leakage from body during the operation	HSE issues	10	Missing factory acceptance test (FAT)	1	10
	Leakage from the seat during the operation	Loss of production	9	Missing FAT and frequency of operation	1	9
Valve body	Damage to the valve body due to pipeline loads	Jeopardizing seal ability and operation of the valve	9	Missing finite element analysis (FEA) on the valve body and internals exposed to the loads and not applying sufficient thickness on the valve body	6	54
Valve seat	Ingress of dense oil in the seat arrangement and malfunction of the valve seat	Jeopardizing seal ability of the valve or galling between the ball and seat	8	Avoid applying valve wash or flushing ports on the valve seat	7	56

(*Continued*)

TABLE 13.8 (CONTINUED)
FMEA Analysis of Pipeline Valves

Valve/ Actuator Component	Potential Failure Mode	Potential Effect of Failure	Severity (S)	Cause of Failure	Occurrence	RPN
Valve seat and balls	Damage to the seat and ball during the commissioning and hydrotest as a result of dirt and welding debris	Failure of ball and seat leads to jeopardizing seal ability of the valve	10	Avoid cleaning the pipeline during commissioning	3	30
Emergency shutdown actuator	Failure in shutting down the emergency shutdown valve during the subsea pipeline failure	Lack of emergency shutdown valve upstream side protection against subsea line failure	9	Missing the function test as well as closure time test for the emergency shutdown valve actuator	1	9

13.4 RESULTS AND DISCUSSION

The last column of Table 13.8 lists the RPNs for five failures associated with pipeline ball valves. The failure rates related to seat components have the highest risk, equal to 95 points. The risk value of 95 is calculated as follows:

The seat failure risk is composed of leakage from the seat during operation due to missing FAT + ingress of hard oil into the seat arrangement due to not washing the valves or flushing the ports on the seats + damage to the seat during hydrotesting and commissioning = 9 + 56 + 30 = 95. Pipeline valves often have a metal seat rather than soft material, as explained in Chapter 1, to facilitate a more robust design. Metal seats are much less likely to be damaged than soft seats due to pressure drop across the valve, particles or metal girt inside the pipeline, and galling caused by contact between the seat and the ball. Hard oil or wax ingress into the valve seat arrangement is the major risk identified and scored for valve seats. According to Chapter 5, this failure item can be avoided by using valve wash or flashing ports on the valve seat. In order to install a valve wash or flushing port, extra holes must be drilled in a valve's body, increasing the chance of leakage. A modular (double block and bleed) valve should be installed on a flushing port to provide double isolation between the seat area and the valve externals. Damage to the valve body caused by pipeline loads is another high failure risk. As explained in Chapter 4, the mitigation approaches are to use sufficient thickness for the valve body as well as FEA.

As a conclusion, this chapter provides a detailed investigation of pipeline ball valves used in export pipelines. FMEA was used to analyze five failure modes related to four pipeline valve components: seat, body, ball, and actuator. According to the analysis, valve seats have an RPN of 95, which is the highest risk score. Ingress of wax (dense oil) into the seat arrangement is the main cause of seat failure with a score of 56 out of 100. This issue can be addressed and mitigated by using flushing ports on the seat arrangement. Having a score of 54 out of 100, damage to the valve body caused by pipeline loads is the second highest failure mode. The main reason for this failure is caused by an insufficient valve thickness calculation and a missing FEA on the valve body. As an alternative to ASME B16.34, an optimal wall thickness calculation based on ASME Sec. VIII Div.02 is proposed to save weight and thickness. Different loads, such as axial, torsion, bending, etc., during design, maintenance, and accident loads should be identified in the valve 3D model for FEA.

QUESTIONS & ANSWERS

1. Find the correct sentences about safety and reliability for pipeline valves.

   A. All pipeline valves are safety critical.
   B. Only emergency shutdown (ESD) valves are safety critical.
   C. Safety integrity level (SIL) analysis is one way to measure the safety and reliability of ESD pipeline valves.
   D. SIL analysis is typically performed in safety analysis report (SAR).
   E. SIL1 provides sufficient safety and reliability for the ESD pipeline valves.
   F. SIL4 provides the highest level of safety as per IEC 61508.

   Answer) Option A is incorrect since only the third pipeline valve after the pig launcher has ESD function and is considered to be a safety-critical valve. Option B is also incorrect since both ESD and PSD valves are safety critical. Options C and D are correct statements. Option E is incorrect since SIL2 or 3 are required for ESD pipeline valves, so SIL1 is insufficient. Option F is right. The correct choices are C, D, and F.

2. Which statement is correct regarding SIL formulas and calculations?

   A. The probability of safe failure equals the sum of detected dangerous and undetected dangerous failure rates.
   B. Diagnostic coverage (DC) is the detected dangerous failure rate divided by the total rate of the failures.
   C. It is possible to correlate SFF to SIL.
   D. In case of a dangerous failure, the safety function goes into its safe state, such as the emergency shutdown of a valve.

   Answer) Option A is incorrect because the probability of safe failure is equal to the sum of the rates of detected and undetected safe failure. Likewise, option B is incorrect since DC is the ratio of the detected rate of dangerous failures divided by the total rate of dangerous failures. The correct answer is C. Option D is incorrect because safe failures put the ESD valve in a safety position.

3. Find the wrong statement about FMEA.

   A. An FMEA is an engineering technique used to identify and eliminate unknown or potential problems and errors in a system.

B. FMEA calculates a risk score before identifying the likelihood and consequences of each failure.
C. If the probability of failure is extremely high and the effect of it is hazardous without warning, the risk score is 100.
D. The risk priority number is equal to severity multiplied by probability of occurrence.

Answer) Options A, C, and D are correct. FMEA calculates risk scores after identifying the likelihood of each failure and its consequences. Therefore, option B is incorrect.

4. What are some scenarios that can lead to pipeline valve failures?

A. If the wall thickness of the valve is too thin to support pipeline loads
B. Solvent injection into seats
C. Constant cycling of the valve which wears out the ball and seat
D. Emergency shutdown of the valve

Answer) Answers A and C are correct. The valve body may be damaged if the thickness of the valve is not sufficient to withstand the load. Cycles of the valve often can cause wear between the ball and seats, which can lead to internal leaks.

BIBLIOGRAPHY

1. Dhillon B.S. (1992). *Maintainability, maintenance, reliability for engineers*, (1st edition). Boca Raton, USA: CRC Press.
2. ESI Technologies Group. (2018). SIL certified ball valves and actuators: A brief review of SIL. [Online]. Available from: https://esitechgroup.com/blog/sil-certified-ball-valves/ [access date: 19th January, 2022].
3. Inst Tools. (2021). Difference between process shutdown and emergency shutdown. [online]. Available from: https://instrumentationtools.com/process-shutdown-and-emergency-shutdown/[access date: 19th January, 2022].
4. International Electrotechnical Commission (IEC). (2010). *IEC 61508, Functional safety of electrical/electronic/programmable electronic safety-related systems, Part 1: General requirements*, (2nd edition). Geneva, Switzerland: IEC.
5. International Electrotechnical Commission (IEC). (2010). *IEC 61508, Functional safety of electrical/electronic/programmable electronic safety-related systems, Part 4: Definitions and abbreviations*, (2nd edition). Geneva, Switzerland: IEC.
6. International Electrotechnical Commission (IEC). (2010). *IEC 61508, Functional safety of electrical/electronic/programmable electronic safety-related systems, Part 2: Requirements for electrical/electronic/programmable electronic safety-related systems*, (2nd edition). Geneva, Switzerland: IEC.
7. International Electrotechnical Commission (IEC). (2010). *IEC 61508, Functional safety of electrical/electronic/programmable electronic safety-related systems, Part 6: Guidelines on the application of IEC 6158-2 and IEC 6158-3*, (2nd edition). Geneva, Switzerland: IEC.
8. International Datalyzer. (2015). *What is FMEA?* [Online]. Available from: https://www.datalyzer.com/knowledge/fmea/ [access date: 21st January 2022].

9. Lean Manufacturing and Six Sigma Definition. (2019). FMEA. [Online]. Available from: http://www.leansixsigmadefinition.com/glossary/fmea/ [access date: 21st January 2022].
10. Sotoodeh K. (2019). Safety integrity level in valves. *Journal of Failure Analysis and Prevention, Springer*, 19, 832–837. https://doi.org/10.1007/s11668-019-00666-2
11. Sotoodeh K. (2021). *A practical guide to piping and valves for the oil and gas industry*, (1st edition). Austin, USA: Elsevier (Gulf Professional Publishing).
12. Sotoodeh K. (2021). *Subsea valves and actuators for the oil and gas industry*, (1st edition). Austin, USA: Elsevier (Gulf Professional Publishing).

Index

A

Actuator, 4, 8, 10, 14, 21, 33, 36, 42, 69, 79, 93, 121–134, 143, 157–158, 166–167, 169–177, 180, 183–184, 190–195, 201–203, 206–214
  Electrical, 8, 14, 36, 123–127, 129, 132–133, 143, 158, 167, 169, 171
  Hydraulic, 8, 14, 36, 122–123, 125, 127–133, 157, 166–167, 169, 171, 184, 201, 203, 206
  Linear, 125
  Scotch and yoke, 124
  Pneumatic, 14, 122–123, 127
  Linear, 125
  Scotch and yoke, 124
American Society for Testing and Materials (ASTM), 19–23, 27–28, 57, 63, 80, 110, 117–118, 160, 206
American Petroleum Institute (API), 7–8, 12, 15, 20, 30, 33, 43, 47, 103, 109–113, 116–118, 136, 140–142, 146–153, 158, 160–161
American Society of Mechanical Engineers (ASME), 2, 12, 19, 24, 30, 50–58, 62, 65, 67–70, 82, 96, 99–111, 117–119, 136–137, 161–162, 184, 199, 213
Antiexplosive decompression (AED), 25
Antistatic spring, 10, 12, 26–28
ATEX, 90

B

Backseat, 75, 139, 147–148, 159–160
Backseat test, 75, 139, 147–148, 159–160
Ball valve, 1–8, 10, 14, 22–23, 26–27, 33–35, 39, 44, 50, 58, 63–68, 71–72, 74, 77, 99–100, 110–115, 124–133, 143, 145–147, 150–151, 154, 161, 170, 176–178, 191–192, 199, 206–207, 212, 214
Basic electric furnace, 80
Bevel end, 104–105, 117, 168–169, 171
Bidirectional seat, 33
Body cavity, 33–35, 141, 146, 153–154
Body test, 140, 141, 150
Bolt, 3–6, 11,12, 18, 22–24, 27–29, 58, 69, 84, 87–90, 99, 101–102, 137, 139, 148, 189–190, 192, 195
  Tightening, 89–90
  Tensioning, 87–88
Bonnet, 3–5, 10–11, 18–23, 27–29, 50, 58, 60, 63, 65–69, 80, 82, 84–90, 99, 109, 116, 131, 139, 147, 154, 186, 190–191, 193, 195–196, 198
Bore, 2, 50, 150
  Full, 2
  Reduced, 2
Break to close (BTC) torque, 129–130, 133
Break to open (BTO) torque, 129–130, 133
Bubble, 143–144, 149–150
Bubble counter, 149
Bubble-tight, 6
Butterfly valve, 2, 15, 77, 102, 118–119
Butt weld, 99, 103–106, 116–117, 119, 136

C

Carbon steel, 18–30, 45–46, 50, 57, 60, 63, 80, 84, 106, 108–110, 117–118, 136, 140, 153, 163, 206
Cast iron, 19
Casting, 79, 80–82, 84, 93–94, 96, 110, 111–116, 118
Carbon dioxide (CO_2), 17, 19, 108–109
Cathodic protection, 28
Cavity relief, 33, 44, 46, 143, 145, 160
Circumferential welding, 105
Client, 2, 8, 12, 38, 79, 91–93, 129, 136, 206
Corrosion, 17–18, 20–30, 60, 71, 92, 99, 101–103, 106–110, 116, 118, 135, 136, 163–171, 189–191
  Chloride stress cracking corrosion (CLSCC), 60
  Corrosion under insulation (CUI), 28
  Internal, 168, 170
  Localized, 28
  Crevice, 17–18, 20, 99, 103, 116, 118
  Galvanic, 17
  Pitting, 17, 60, 136
Construction yard, 25, 29, 43, 92–93, 106, 110, 116, 118, 138, 163, 168, 170, 175
Corrosion resistant alloy (CRA), 17, 106, 108
Coating
Compact flange, 100–101, 119
Closure member, 10, 20, 26, 33, 38, 115, 118, 129, 132, 141, 145–146, 151, 155, 159–160, 166
Closure test, 139, 141–144, 146–147, 149–150, 158–161
Cost, 2, 4, 6, 13, 17, 35–36, 44, 49–50, 64, 91, 94, 99, 101–102, 109, 114, 122, 135, 150, 158, 170, 175
  Capital cost, 94
  Operational (OPEX), 158
Crate, 173
Cross-section drawing, 9–10, 27, 92, 187
Cutting tool, 84–85, 95
Cylindrical nut, 65–69

D

Dangerous detected failure, 203–204, 206
Dangerous failure, 202, 204, 213
Dangerous undetected failure, 202–203, 206
Delivery time, 1–2, 4, 13
Design life, 17
Det Norske Veritas (DNV), 2, 109, 112, 128
Diagnostic coverage, 204, 213
Disk, 3–4, 20, 26, 150
Double-acting actuator, 124–125, 130–131, 157
Double block and bleed (DBB), 36, 44–45
Double isolation and bleed (DIB), 33, 35–36, 38–39, 41–46, 77, 138, 141, 146–147, 159, 184
Double piston effect (DPE), 33–36, 38–45, 77, 146–147, 159
Drain plug, 10–11, 27
Drift test, 139, 150–151, 160
D-ring, 177, 179–180

E

Electrical actuator, 8, 14, 36, 123–127, 129, 132–133, 143, 158, 167, 169, 171
Emergency shutdown function, 8, 45, 132, 201
Emergency shutdown valve, 124, 143, 163, 212–213
Break to close (BTC) torque, 129–130, 133
Break to open (BTO) torque, 129–130, 133
Export pipeline, 1–2, 19, 71, 106–107, 109, 111, 117, 126, 128, 132, 206–207, 212

F

Factory acceptance test (FAT), 4, 91, 94, 135, 162–163, 211
Fail safe close (FSC), 124, 126–127, 134
Failure mode and effect analysis (FMEA), 209–215
Final inspection, 79, 92, 163
Finite element analysis (FEA), 38–39, 41, 58, 61, 63–65, 69, 211–213
Fire box, 130–131, 134
Fire safe design, 10, 152
Fire safe ring, 11
Fire test, 139, 151–152, 154–155, 160
Flange, 4–7, 10, 14, 50, 70, 72, 90, 92, 96, 99–102, 105, 116, 188–119, 126, 136, 138, 145, 148, 162–165, 186, 188, 189–193, 198
  Connection, 4–5, 10, 14, 105, 191
  Face, 163–164, 189–190
  Ring type joint (RTJ), 189–190
Floating ball, 7, 35
Fluid, 2, 5, 7, 10, 17, 19–20, 21, 24, 33–34, 36, 38–39, 45–46, 50, 74–77, 91, 101, 103–104, 116, 123–124, 135–139, 143–144, 146–147, 149, 150, 158–159, 166, 184, 188, 190, 192, 201
  Flow, 2, 17, 33, 38–39, 45, 116, 166, 201
  Leakage, 5, 20, 33, 36
  Pressure, 7, 33, 75, 136, 139
Flushing fluid, 75–77
Flushing port, 71–77, 141, 186, 189–190, 198, 211–213
Flushing procedure, 75
Forging, 22, 24, 79–80, 82, 94
Foundry, 80, 86
Foundry sketch, 80

G

Galvanic corrosion, 17
Galvanic effect, 28
Galvanized, 27
  Bolt, 27
  Nut, 27
Gasket, 12, 18, 20, 27, 72, 87, 90, 99, 101, 165, 186, 189–192
Gate valve, 2, 26, 35, 46, 77, 129, 133, 147–148
Gearbox, 8, 10, 121, 123, 127, 192
Gland, 148
Gland flange, 148
Graphite, 11–12, 25, 27–28

H

Handwheel, 8, 121, 126–127, 192–193, 198
Handling, 173–181
Hard oil, 212
Health, safety and environment (HSE), 6, 91, 94, 128, 135, 209
High-pressure class, 8, 12, 36, 99–100, 123, 128
Hot dip galvanized (HDG), 24, 27–28
Hydrogen sulfide (H_SS), 19, 23–24
Hydraulic, 8, 14, 21, 36, 88–89, 121–134, 157, 166–169, 171, 174, 176–178, 184, 201–203, 206
  Actuator, 8, 14, 36, 122–123, 125, 127–133, 157, 166–167, 169, 171, 184, 201, 203, 206
  Fluid, 123–124, 167
  Force, 88
  Power, 121–123, 131
Hydrostatic seat test, 132, 158
Hydrostatic shell test, 140–141, 158–160

I

Inconel, 20–21, 27–29, 84, 99, 111–112, 116, 118
Inhibitor, 27–28, 60, 136, 163, 165–166, 168, 171
  Corrosion, 27–28, 60, 136, 163, 165–166, 168, 171

Inline maintenance, 4, 13–14
Inspection, 5, 12, 71, 79, 81, 92–93, 96, 106, 114, 116, 136–137, 150, 163, 183, 186
Inspection and test plan (ITP), 79, 93, 96
Installation, operation, and maintenance manual (IOM), 137, 189–191
Instrumentation, 43
Insulation, 28
International Organization of Standardization (ISO), 33, 47, 119, 136, 143–144, 146, 150–151, 158, 161–162

J

Johan Sverdrup, 127–128
Joint efficiency, 105–107
Junction box, 166–167, 171

K

Kick-off meeting, 79

L

Lashing chain, 177–181
Leakage rate, 143–150, 159, 161, 186, 192
Low-alloy steel, 17–18, 20, 23–24, 27–29, 106, 108–111
Lifting, 4, 50, 83–84, 92–95, 170, 173–181
Lifting lug, 83–84, 93, 95, 173–176, 181, 193, 198
Linear actuator, 125, 127, 133
  Diaphragm actuator, 125
  Piston actuator, 125
Lip seal, 11–12, 25, 27–28, 30, 33, 75
Liquid penetration test, 114–115

M

Machining, 79, 82, 84, 86–87, 93–94, 96, 122
  Drilling, 84
  Milling, 84
  Turning, 84
Maintenance, 2–5, 13–14, 36–38, 45–46, 59–61, 63–65, 69, 71–72, 91, 99, 101, 135, 137, 146, 150, 183–199, 206, 213–214
Magnetic particle inspection, 81, 116
Manual valve, 121
Manufacturing process, 79–97
Marine Environment, 29
Marking, 43, 45–46, 50, 90, 92
Mechanical joint, 99, 101
  Clamp, 99, 116, 138–139
  Hub, 99–100, 116, 138–139
Metal-to-metal, 26
Modular valve, 74–75, 77, 186, 189–190

N

Non-destructive testing (NDT), 79, 93, 114
NORSOK, 64, 70, 80, 96
Norwegian offshore, 99, 104
Nut, 3–4, 11–12, 18, 22–24, 27–29, 50, 65–67, 69–70, 84, 87–89, 99, 127, 139, 148, 189–192

O

Offshore pipeline, 1, 2, 7, 9, 11, 12, 13, 18–22, 24–26, 29–30, 36, 50, 61–62, 64–65, 72, 99, 109, 135, 163
Offshore platform, 1, 49, 106, 135, 209

P

Packing, 79, 92–93, 139, 148, 163, 170, 183–184, 197
Painting, 92, 163
Particle-containing services, 7, 76
Part list, 9, 11–12
Particles ingress in seats, 76
Passive fire protection (PFP), 131
Pipeline valve, 1–215
Piping injected gadget (PIG), 2–3, 7, 12–13, 36–39, 45, 71, 106, 109, 112, 126–128, 132–133, 150, 213
Platform, 1–2, 24, 29, 49, 61, 70, 104, 106, 109, 128, 135, 209
Pneumatic actuator, 14, 122–123, 127
Polyether ether ketone (PEEK), 11
Post-weld heat treatment (PWHT), 82, 110–111, 118
Preservation, 79, 92–94, 163, 166, 168–171, 183
  External, 168
  General, 163
  Internal, 166
Pressure nominal (PN), 6, 39, 50, 99
Pressure safety valve (PSV), 146
Pressure test, 59, 86, 91–92, 94, 135–139, 153, 157–158, 160, 163–164, 171
Pup piece, 6, 10–11, 20, 24–25, 27, 29–30, 65, 83, 93, 99, 110–111

Q

Quality control, 79, 81–82, 86, 116
Quarter-turn, 10, 33, 122, 124–125

R

Raw material, 80
Radiography test, 106, 115, 118
Reliability, 35–36, 44–45, 94, 101, 114, 121, 160, 162, 185, 201–211

Repair, 3, 11, 72, 82, 93, 116, 118, 185–192, 197–198, 204, 206
Riser valves, 1
Root face, 104, 117
Running torque, 129, 133

S

Safe failure, 202, 205–207, 210
Safe failure fraction (SFF), 208, 208–209
Safety, 6, 15, 31, 35–36, 38, 45, 64, 71, 75, 77, 90–91, 94, 101, 107, 121, 123, 128, 131, 135, 137, 149–150, 152, 160, 173, 176, 185, 201–215
Safety datasheet, 75
Safety factor, 129–130, 133
Safety integrity level (SIL), 201–202, 213
Safety relief valve, 141
Safety valve, 35, 141, 146
Seal, 5–6, 12, 18, 20, 25, 27–28, 30
Sealant injection, 72–76, 141, 185, 197
Sealant injection fitting, 11, 72–76, 185–186, 188, 192, 197
Sealing, 11, 17, 26, 28, 33, 71, 101, 110, 116, 141, 151, 159–160, 186, 188–190, 192
  Antiexplosive decompression (AED), 25
Seat, 2, 6–7, 10–12, 18, 20–29, 33–45, 58, 64–69, 71–77, 80, 82, 87, 91, 115–116, 118, 129, 132, 136–137, 139–142, 155, 158–162, 183–198
  Metallic, 7, 22, 76, 143, 146, 151, 158, 162
  Soft, 6–7, 212
Seat scraper, 28, 71–72, 76
Self-relieving (SR), 33–34, 143
Shaft, 10, 124, 151, 186
Shell, 92, 137, 140, 148
Shell test, 91, 139, 140, 141, 149–150, 157–160
Shipment, 92, 96
Side-entry, 3–4, 6, 13–14, 64
Single acting actuator, 124–125
Single piston effect (SPE), 33, 82
Slab gate valve, 35, 147–148
Socket weld, 103–104
Solenoid valve, 166–167, 171
Special tool, 180–181, 192, 195–196, 199
Split body, 4
Spring guide, 10
Stainless steel, 18, 43, 45, 50, 60, 110, 119
  Austenitic, 26, 60
  Duplex, 24–25, 29, 104–106, 108–111, 116–119
  Inconel, 20, 21, 27–29, 84, 99, 111–112, 116, 118
  Inconel, 625, 20–21, 27, 29, 84, 99, 111–112, 116, 118
  Inconel X750, 27–28
  Martensitic, 20–22
  17–4PH, 22, 29
  Super duplex, 119
Standards, 33, 64, 68, 93, 99, 105, 109, 112, 136, 137, 151, 155, 202
  API, 7–8, 12, 15, 20, 30, 33, 43, 47, 103, 109–113, 116–118, 136, 140–142, 146–153, 158, 160–161
  ASME, 2, 12, 19, 24, 30, 50–58, 62, 65, 67–70, 82, 96, 99–111, 117–119, 136–137, 161–162, 184, 199, 213
  ASTM, 19–23, 27–28, 57, 63, 80, 110, 117–118, 160, 206
  DNV, 2, 109, 112, 128
  ISO, 33, 47, 119, 136, 143–144, 146, 150–151, 158, 161–162
  NORSOK, 64, 70, 80, 96, 100, 108, 119
Stem, 4, 10–11, 27
Stem bearing, 12, 26, 28, 30
Stem enlargement, 4
Stem key, 10–11, 27
Stem sealing, 11
Stopper plate, 177, 179–180
Stress, 62, 63
  Bending, 62, 63
  Bending plus membrane stress, 62, 63
  Peak, 62, 63
Stud bolt, 11–12, 27, 89, 99, 189, 190, 192

T

Tag plate, 45–46, 90, 184
Tensile, 62, 63, 80, 82
  Strength, 63
  Stress, 62
Test fluid, 136–137, 143–144, 149–150, 159
Testing, 12, 19, 50, 58–59, 61–65, 79–80, 82, 91–93, 99, 114, 135–164, 170, 207, 212
Test preparation, 136
Test pressure, 58, 60, 91, 136–137, 140, 143, 149–150, 153, 157–160
Test procedure, 136
Threaded connection, 99, 102
Thermal spray aluminum (TSA), 28
Through conduit gate (TCG) valve, 2, 26, 33, 125, 133, 139
Thrust washer, 12, 26, 28, 30
Titanium, 18–19
Top-entry, 2, 4–6, 13–14, 18, 64, 170
Topside, 49–50, 61, 69–70, 102, 104, 128
Torque, 33, 35–36, 41–42, 87–88, 90, 122–123, 129, 130–133, 137, 139, 143, 158–160, 189–190
Torque tool, 87–88, 90
Transition piece, 6, 18, 20, 24–25, 28–29, 58, 83–84, 110–117
Trim, 20, 109

Trunnion, 7, 35
Trunnion-mounted, 7, 35
Trunnion-mounted ball, 7, 35

U

Ultrasonic examination, 114, 116, 118
Unidirectional, 33, 146, 184
Upstream, 2, 7, 19, 38, 111, 128, 147, 150, 212

V

Vacuum degassing process, 80
Valve assembly, 95, 110, 137, 173
Valve internal, 2, 4, 20, 110, 118, 154, 184
Vent plug, 11, 27
VCI, 166, 168, 171
Visual inspection, 81, 116, 186

W

Wafer, 101–103
Wall thickness, 24, 50, 55, 57–58, 64–69, 106–108, 117, 214
Wall thickness calculation, 50, 56–57, 67, 107–108, 213
Wall thickness validation, 58, 69
Wax, 2, 71–72, 75–77, 164, 190, 212–213
Weight reduction, 50–70, 106, 129
Welded joint, 4–5, 104, 119
Weld joint efficiency, 105–106
Welding, 4–6, 15, 17, 24, 29–30, 50, 71, 76, 79, 82–84, 93, 95–99, 119, 136, 138, 168, 183–184, 197, 199, 212
Weld overlay, 27, 99
Wooden box, 170, 173, 183

Y

Yard, 6, 25, 29, 43, 92–93, 106, 110, 116, 118, 138, 163, 168, 170, 175

Z

Zero leakage, 6